MANAGING PROCESS INNOVATION

From Idea Generation to Implementation

Series on Technology Management*

Series Editor: J. Tidd (Univ. of Sussex, UK)   ISSN 0219-9823

Published

*For the complete list of titles in this series, please write to the Publisher.

SERIES ON TECHNOLOGY MANAGEMENT – VOL. 17

MANAGING PROCESS INNOVATION

From Idea Generation to Implementation

Thomas Lager

Grenoble Ecole de Management, France

Imperial College Press

ICP

Published by

Imperial College Press
57 Shelton Street
Covent Garden
London WC2H 9HE

Distributed by

World Scientific Publishing Co. Pte. Ltd.
5 Toh Tuck Link, Singapore 596224
USA office: 27 Warren Street, Suite 401-402, Hackensack, NJ 07601
UK office: 57 Shelton Street, Covent Garden, London WC2H 9HE

British Library Cataloguing-in-Publication Data
A catalogue record for this book is available from the British Library.

Cover Photo: Courtesy of Linde AG
Linde is the only company, world-wide, that designs, owns and operates hydrogen and synthesis gas
plants using its own technology. Experience from its own plant operations flows into plant construction.
The picture shows a Linde Steamreformer in Brunsbuettel, Germany. Hydrogen and carbonmonoxide
are transported to customers directly through a pipeline.

Series on Technology Management — Vol. 17
MANAGING PROCESS INNOVATION
From Idea Generation to Implementation

ISBN-13 978-1-84816-605-9
ISBN-10 1-84816-605-2

Typeset by Stallion Press
Email: enquiries@stallionpress.com

Printed in Singapore.

Dedication

This book is dedicated to Gunilla Bergdahl, my research assistant and professional co-ordinator and a passionate supporter of my ambition and constant struggle to deliver research results that are both rigorous and provide practical, relevant advice to industry professionals: a true comrade-in-arms.

Acknowledgements

I would first of all like to thank Professor Joe Tidd for his belief that a book on this subject would be important. Many thanks also to him for taking the time to read an early manuscript and for giving important advice and guidance in the writing process.

Two readers of the first, rather indigestible draft provided important comments from an industrial perspective: Emilia Liiri-Brodén, Development Manager of Holmen Paper AB and Dr Stefan Gustafsson, R&D Manager at Höganäs. Many thanks for important suggestions and for giving a first indication and feedback that the manuscript had the potential to become useful even to experienced industry professionals.

In the initial stages of this book, my mentor Professor Derrick Ball provided helpful comments and advice for which I am very grateful. For suggestions and comments on the manuscript I also sincerely thank Professor Sylvie Blanco from Grenoble École de Management, Dr Johan Frishammar from Luleå University of Technology and Dr Peter Samuelsson, Vice President R&D at Outokumpu for taking the time to read selected parts of the book in spite of their tight agendas. Many thanks to Alan Gilderson, who has helped to transform my sometimes ugly English prose into a more digestible manuscript.

I also thank two of my previous doctoral candidates, Dr Markus Bergfors of Conchius (Shanghai) Ltd (2007) and Dr Andreas Larsson of Vinnova (2007) for permission to use some material from their doctoral theses. For permission to use a large part of a joint forthcoming publication in Chapter 7 and for reading Part 3 of the book, I especially thank you, Markus.

Some material from previously published or forthcoming publications in scientific journals and presentations has been used in this book, sometimes as large extracts of text and even as the main part of individual chapters. My sincere thanks for permission to use this kind of material go to:

Professor Joe Tidd, editor of International Journal of Innovation Management (IJIM);

Dr Jeff Butler, editor of R&D Management;

Professor Abby Ghobadian and Professor David Gallear, editors of International Journal of Process Management and Benchmarking (IJPMB).

Some material has come from my publications, forthcoming publications and parts of papers still under review. Many thanks to my co-authors:

Dan Hallberg and Pierre Eriksson (LKAB) for permission to use material in Chapter 9, from the joint publication in IJIM (Lager *et al.*, 2010);

Dr Johan Frishammar (Luleå University of Technology) for permission to use material in Chapter 10, from a joint presentation at an R&D management conference (Lager and Frishammar, 2009);

Dr Per Storm (Raw Materials Group) and Ulf Holmqvist (Höganäs) for permission to use material in Chapter 11, from a joint publication under review.

Introductory Quotation

"The greatest risk in innovation is not to take any risks at all!"

Hans Rausing: innovator, entrepreneur, industrialist and former
major owner of the Tetra Pak International SA.

Preface

Is there a need for one more book on innovation? In the fourth edition
of the important and voluminous "*Strategic Management of Technology
and Innovation*" by Burgelman, Christensen and Wheelwright (2004),
for example, there is hardly anything at all about the innovation of
process technology. The phenomenon is not uncommon, and there are
few publications dealing with process innovation at company level. One
possible explanation for this state of affairs is that the subject is felt to
be too complicated for researchers lacking a technological background
and, particularly, knowledge of the industrial environment in the
process industries.

Process industries and other manufacturing industries: Are they different?

There are a number of similarities between the process industries and
other manufacturing industries with regard to development activities
and how they are performed, but there are also a number of differences.
The main difference is probably the strong relation between product
development and the development of process technology. An understanding of production process technology is thus essential to the
development of new products. The automotive industry has recognised
the importance of "design for manufacturability" (Boothroyd *et al.*,
1994). The same concept is also valid in the process industries, but in
this book it will be called "innovation for processability." The concept
of "dominant design", describing a product that outperforms its

competitors, was introduced by Utterback (1994, p. 18). This concept name does not fit products commonly produced in the process industries very well, since they are not designed assembled products. The idea behind the concept is however still relevant and valid, and here "dominant performance" is introduced and will be used to describe the "best-in-class" products in the process industries. The following list also highlights some other differences between innovation in the process industries and that in other manufacturing industries:

- *Development work is done in a laboratory/pilot plant/production plant environment and not in a design office.*
- *Constructing prototypes is not an intermediate development stage in product development but is replaced by test runs in pilot plants or full production, manufacturing batches of new products for customers and verifying process conditions.*
- *Customers are often industrial business-to-business or B2B in long supply chains, before finally reaching the end user business-to-consumer or B2C.*
- *Process development often takes place in collaboration with manufacturers of process equipment and suppliers of raw material and reagents.*
- *Changes in the company's portfolio of products and product varieties are often very complex, since the company's production structures are long internal interconnected production chains.*
- *Final quality of products and intermediate product properties are related to available raw material properties.*

The conclusion is that the differences in the development context motivate a separate study of product and process innovation in the process industries.

Process innovation is not, however, the only area of innovation that does not receive proper attention. Considering that the process industries make up a large part of all manufacturing industries, the management of innovation and technology in the process industries is an area that deserves more recognition in academic research and in other publications. In this book it has sometimes been necessary to discuss process innovation from the perspective of all kinds of innovation and innovation-related activities in the process industries in order to give a better understanding. Readers interested solely in process innovation are recommended to read those other parts extensively.

Process innovation (development), as defined in this book and distinct from product innovation, is in itself a strategic area of innovation for most firms in the process industries seeking to secure competitive and cost-efficient production technology. The supportive development of process technology for successful product development is by no means considered to be of less importance. This work is, however, an activity classified as part of product innovation as defined here, even if carried out by the organisation for process innovation.

Readership

Research on the management of process innovation can be included in the larger research area often referred to as "research on research" and as such must be considered an applied part of this research area. Because of that, and because the readership for this book includes both academic scholars and industry professionals, the aspects of company usability (relevance) are stressed — maybe even sometimes at the expense of some academic rigour. The need for such a focus has been well demonstrated by Ball (1998), and the aim and overall purpose of this book are thus:

> To introduce existing and improved concepts, models and management tools in the area of process innovation in order to improve the performance of innovation at the company level. Furthermore, to provide theoretical insight into the subject matter and to give ideas for further research in the area of process innovation and innovation in general in the process industries.

A note to the reader

An oft-quoted statistical argumentation attributed to Samuel Johnson states: "You do not have to eat the whole ox to know that the meat is tough." The phrase is instructive for scholars in statistics, but unfortunately we all know that different parts of the ox really do differ in tenderness, and that people have different preferences for which parts to eat. Different parts of this book may be of different tenderness, more or less tough to digest, and furthermore different parts will probably interest different readers. For the reader to find out which parts are of special interest to him, I am sorry that I must advise him to "eat the whole ox."

Please try to take a slice of the ox at a time and do not hesitate to start with the part that interests you most. It is thus possible to start with Part 5, Process Innovation Performance, or, if you are currently busy working out how to develop R&D work processes, please feel free to start with Part 3. However, since the book does have an intended structure for improved understanding, I sincerely recommend that you start from the beginning of the story. Please observe that further references are given at the end of each part of the book.

References

Anderson, P and Tushman, ML (1990). Technological discontinuities and dominant designs: A cyclical model of technological change. *Administrative Science Quarterly*, 35, 604–633.

Aylen, J (2010) Open versus closed innovation: Development of the wide strip mill for steel in the USA during the 1920's. *R&D Management*, 40.

Ball, DF (1998). The needs of R&D professionals in their first and second managerial appointments: Are they being met? *R&D Management*, 44, 139–145.

Bergfors, M (2007). Designing R&D organisations in process industry. Doctoral dissertation, Luleå University of Technology, Dept of Business Administration and Social Sciences.

Bergfors, ME and Larsson, A (2009). Product and process innovation in process industry: A new perspective on development. *Journal of Strategy and Management*, 2, 261–276.

Bergfors, M and Lager, T (2011). Innovation of process technology: Exploring determinants for organizational design. *International Journal of Innovation Management*, February forthcoming.

Boothroyd, G, Dewhurst, P and Knight, W (1994). *Product Design for Manufacture and Assembly*. New York: Marcel Dekker Inc.

Burgelman, RA, Christensen, CM and Wheelwright, SC (2004). *Strategic Management of Technology and Innovation*. New York: McGraw-Hill/Irwin.

Drucker, P (1993). *Innovation and Entrepreneurship*. New York: Collins.

Drucker, P (1998) *On the Profession of Management*, Boston, Harvard Business School Press.

Foster, RN (1986). *Innovation — The Attackers Advantage*. New York: Summit Books.

Grönlund, J, Rönnberg Sjödin, D and Frishammar, J (2010). Open innovation and the stage-gate process: A revised model for new product development. *California Management Review*, 52, 106–132.

Hein, P (1985). Att formulera ett problem är att lösa det (in Swedish). *Forskning och Framsteg*, 6, 47–50.

Johnson, MW, Christensen, CM and Kagermann, H (2008). Reinventing your business model. *Harvard Business Review*, December, 51–59.

Lager, T (2002). A structural analysis of process development in process industry — A new classification system for strategic project selection and portfolio balancing. *R&D Management*, 32, 87–95.

Lager, T (2005). The industrial usability of quality function deployment: A literature review and synthesis on a meta-level. *R&D Management*, 35, 409–426.

Lager, T (2008). Using Multiple Progression QFD for Roadmapping Product and Process Related R&D in the Process Industries. *14th International Symposium on Quality Function Deployment*, Beijing, China.

Lager, T and Beesley, N (2010). Start up of new process technology in the Process Industries: Organizing for an extreme event. *2nd iNTeg-Risk Conference: New Technologies & Emerging Risks*. Stuttgart, European Virtual Institute for Integrated Risk Management.

Lager, T and Blanco, S (2010). The Commodity Battle: A Product-Market Perspective on Innovation Resource Allocation in the Process Industries. *The R&D Management Conference*. Manchester.

Lager, T and Frishammar, J (2009). Collaborative development of new process technology/equipment in the process industries: In search of enhanced innovation performance. In *R&D Management Conference*, J Butler (ed.), Vienna.

Lager, T and Frishammar, J (2010). Equipment supplier/user collaboration in the process industries: In search of enhanced operating performance. *Journal for Manufacturing and Technology Management*, 21.

Lager, T, Hallberg, D and Eriksson, P (2010). Developing a process innovation work process: The LKAB experience. *International Journal of Innovation Management*, 14, 285–306.

Larsson, A (2006). Innovation and technology in process industry: A process management perspective on technology strategic planning. *International Journal of Process Management & Benchmarking*, 1, 201–219.

Larsson, A (2007). Strategic management of intrafirm R&D in process industry. Doctoral dissertation, Luleå University of Technology, Dept of Business Administration and Social Sciences.

Larsson, A and Bergfors, M (2006). Heads or tails in innovation strategy formulation? Porterian or Penrosian, letcontext determine. *International Journal of Process Management & Benchmarking*, 1, 297–313.

Larsson, A and Sugasawa, Y (2007). A unified methodology for innovation strategy formulation: Product and resource perspectives on the firm combined. *International Journal of Process Management & Benchmarking*, 2, 118–137.

Miyashita, K and Russel, D (1995). *Keiretsu: Inside the Hidden Japanese Conglomerates*. New York: McGraw-Hill.

Skinner, W (1992). The hareholder's delight: Companies that achieve competitive advantage from process innovation. *International Journal of Technology Management*, 41–48.

Storm, P, Lager, T and Samuelsson, P (2010). A material conversion perspective on firm innovation in the Process Industries: The production system revisited. *The R&D Management Conference*. Manchester.

Utterback, JM (1994). *Mastering the Dynamics of Innovation: How Companies Can Seize Opportunities in the Face of Technological Change*. Boston, MA: Harvard Business School Press.

About the Author

Thomas Lager is an affiliated professor at the Centre for Innovation, Technology and Entrepreneurship at the Grenoble Ecole de Management in France. He was previously the founder, adjunct professor and director of the Centre for Management of Innovation and Technology in Process Industry (Promote) at the Luleå University of Technology in Sweden. He is also Managing Director of blinab, a "boutique" management consulting firm in France. He holds an MS degree in Mining Engineering from the Royal Institute of Technology, Sweden. He has a PhD in Mineral Processing and a PhD in Business Administration and Economics from the Luleå University of Technology, Luleå. Dr Lager has also served 15 years in the process industry, mainly in the capacities of production engineer in Sweden and Africa, and R&D manager in the Swedish mining industry.

Academic affiliation

Thomas Lager,
Centre for Innovation, Technology and Entrepreneurship,
Grenoble Ecole de Management,
BP 127 — F-38003 Grenoble Cedex 01, France.
E-mail: thomas.lager@grenoble-em.com
Web: www.grenoble-em.com

Company affiliation

Dr. Thomas Lager
blinab
454, Route d'Uriage appt. 241
38410 St Marin d'Uriage
France

Contents

PART 1

INNOVATION IN THE PROCESS INDUSTRIES

One way to improve industrial innovation is by better management. This book is thus about management of process innovation and technology, and not about technology as such. We must never forget that knowledge of innovation management can be acquired as "learning by doing" in the industrial environment, and that books or other scientific publications must be regarded as supplementary, albeit important, sources of learning. The more theoretical and scientific publications can thus serve as a framework for industry professionals and also provide structure for their own practical experience. The discipline of innovation management is however not an old one, and there are still areas that are not yet widely explored and where good theories are still lacking. Management of process innovation can be included in this group.

The first part of the book presents the process industries as a group of several industry sectors — a family of industries often sharing common attributes, particularly in the production process. Recent evidence from the 2000 most innovation-intensive and largest companies worldwide and in Europe, representing about 80 percent of worldwide innovation, shows that about 30 percent of them can be included in the process industries cluster. Innovation in this large area of total manufacturing industry is of great concern, not only to industry professionals and academics, but to society at large. However, research on innovation management and operation management in this cluster of industries at the company level is still in its embryonic stage compared to such research at the industry or society level.

Innovation at the process industries level has not received its fair share of research in the area of management of innovation and technology, either. Although one can estimate from previous research that innovation in many sectors of the process industries claims only a surprisingly small share of all a company's resources, good management of innovation in this area is however very important in terms of securing an efficient production process and competitive products for the future on the world market.

Chapter 1

Introduction

"On the one hand we find descriptive models which merely answer the
question, 'why are we the way we are?' The manager who learns from
a researcher that his organisation has the characteristics of an adhocracy,
and that this can be easily explained, will in most cases merely take note
of this announcement, and just think: 'So what...?' Normative ideas
and models, on the other hand, provide a direction towards which an
organisation has to proceed in order to innovate successfully."

Jan Cobbenhagen *et al.* (1990)

Does insight from research on Management of Innovation and
Technology (MOT) provide new conceptual tools and models that
give guidance to industry professionals and opportunities for them to
improve their innovation management practices? Or does management research merely structure, learn and explain good management
behaviour in order to give other not-so-successful firms a helping
hand (Cobbenhagen *et al.*, 1990)? The question is related to how
good management research ought to be carried out in the future in
order to create a win-win situation for both academia and industry.
The author's perspective on good research is much in line with the
view advanced by Kurt Lewin (1946) and further promoted by Chris
Argyris (2002), as the approach of active involvement combined with
expected insights developed through research.

Using the definition of "management innovation" as "the generation and implementation of a management practice, process, structure,

3

or technique that is new to the state of the art and is intended to further organisational goals (Birkinshaw *et al.*, 2008)", the conclusion is that good interactive behaviour between academia and industry is considered a fruitful approach. Such a procedure and ambition is then very close to what Kaplan (1998) defines as "innovation action research", and as such emphasizes the close ties to developing a new management theory that is both conceptually sound and generally applicable. Some of the research results presented in this book are the outcome of such an approach; some are the outcomes of more conventional research and a large European survey of R&D managers.

In the first section of this introduction you will find a theoretical framework that aims at picturing different aspects of innovation in general. This structure and content are then not only integrated in the book but also intended to provide food for thought for both scholars and industry professionals. In the following section you will find some publications dealing with innovation in the process industries. These are intended not only to present options for research design, but also to provide a list of some important contributions and early publications in this area. The discipline of management of innovation and technology is fairly young, and there are consequently many areas where solid empirically verified theories are lacking. In a following section you will therefore find the author's somewhat crude personal impression of this state of affairs and an articulated need for such improved theoretical knowledge. Finally the use, or perhaps misuse, of some selected innovation concepts is discussed in order to excuse the appearance of many similar concepts which it has often been found necessary to employ for various reasons.

1.1 A theoretical platform

Findings from research in the area of innovation management are unfortunately sometimes presented in such a manner that it is difficult to understand whether the results are what a company should aim at (normative or prescriptive), or whether they simply picture the existing good or poor state of affairs (descriptive). Going from descriptive research to prescriptive research will make results more interesting to industry professionals, but the bottom line must nevertheless be that the presenter of a study must make it clear whether the findings are to be regarded as simply descriptive, or whether they are prescriptive and thus propose possible ways to improve a company's innovation behaviour.

A framework for research can be described in different ways. Miles & Huberman (1994, p. 18), for example, speak of intellectual bins:

> "Bins come from theory and experience and (often) from the general objectives of the study envisioned. Setting out bins, naming them, and getting clearer about their interrelationships lead you to a conceptual framework. ... Frameworks can be rudimentary or elaborate, theory-driven or commonsensical, descriptive or causal."

The framework and platform for this book have emerged gradually during the author's industrial and academic life, and the platform rests on four cornerstones that each relate to different aspects of innovation in the process industries. The following cornerstones can be regarded as four theoretical propositions that have guided the presentation in this book, but could possibly also be reflected upon and used in future studies of innovation in the process industries:

A contextual aspect

A company's internal and external innovation environment and its external business and competitive environment should guide its innovation strategy and programmes. The context for development thus differs from one company to another, and innovation activities must be adapted accordingly, in line with contingency theory (Woodward, 1965). Since these environments are also rapidly changing, development activities must consequently adapt to those changes in a dynamic manner.

A structural aspect

There are often many categories of development activities going on in a company. There is therefore a need to properly define and categorize those development activities in order to understand the nature of each development activity (Lager, 2001). This will give a better awareness of the development activity's economic and strategic potential, need for resources and skills in the organization, risks, and demands on the individuals involved.

A processual aspect

In all company innovation processes there is a proper balance between order and chaos that may change over time in the company and in an individual innovation project. Each firm's innovation processes must find this proper balance that fits the company, the development activity, the individual innovation project and the individuals involved (Norling, 1997).

A relational aspect

Many of the individual parts of a company's innovation activities are related to each other. One must not only understand and optimize the individual parts on their own, but also understand how they relate to and interact with each other and optimize the whole. This interaction within the firm's unique innovation culture makes it necessary to take a holistic view of company innovation.

The four cornerstones are in many ways interdependent. The first cornerstone stresses the importance of the rapidly changing internal and external company environment and the need for the R&D organization to adapt to changing circumstances. The innovation processes mentioned in the third cornerstone must be adapted to this internal and external innovation environment. The development processes must not only be adapted to the company-specific environment, but must also be changed when the circumstances change.

The second cornerstone emphasizes the importance of distinguishing more clearly between different categories of innovation activities, because different categories require different working practices, skills and resources. This aspect certainly influences the third cornerstone, which refers to the different types of innovation processes covering a spectrum from total lack of structure or evolving structure to highly formalized and plannable processes or parts of work processes.

The last cornerstone addresses the interdependence between different innovation activities and practices as well as total integration within the company. The first cornerstone favours research results from case studies because they are more context-related and thus often more understandable. Behaviour or performance in a specific case may thus be more easily transferred or applied in another company. Some chapters of this book thus rely to a large extent on single case studies, while the main body of empirical evidence is based on a large survey to European process industries (Lager, 2001) or from statistical analysis of information from a large European database (Guevara *et al.*, 2008).

1.2 In search of better theories and best practice for management of innovation and technology in the process industries

Criteria for the selection of clusters of firms or industries to study in management research could be size, geographical location, types of production process, type of products, type of ownership, high tech or low tech, mature industries versus emerging industries, etc. It is thus relevant to ask whether innovation in the cluster of industries here called the process industries is of interest for study and discussion.

All possible comparisons between clusters of industries can indeed be used as long as they serve the purpose of illustrating a phenomenon or issue of interest — in this case, for studying Management of innovation and Technology (MOT) in the process industries. In studies of organizational behaviour, there are probably also a number of areas where the process industries can learn from other manufacturing industry. If for example efficient product distribution systems are the object of a study, it may very well be advantageous for a chemical company to compare itself with companies outside of the process industries.

One must however be careful when comparing firms and industries that differ too much in their internal and external operating environment, since this may give rise to problems and even errors in the interpretation of the results and the further use of the results for industry professionals. A number of studies and books on innovation in the process industries have used different selections of samples:

- Some studies and books have dealt with samples of manufacturing industry including companies from various sectors of the process industries (e.g. (Allen, 1967; Booz Allen and Hamilton, 1982; Chakrabarti, 1988; de Margerie, 2009; Freeman and Soete, 1974; Kline, 1985; Langrish *et al.*, 1972; Schroeder *et al.*, 1986; Utterback, 1994)
- Others have studied one or more segments (sectors) of the process industries such as the chemical industry (Achilladelis *et al.*, 1990; Aftalon, 2001; Freeman *et al.*, 1963); the petrochemical industry (Achilladelis, 1975; Enos, 1962; Spitz, 1988; Stobaugh, 1988) and the pharmaceutical industry (Pisano, 1997).
- Yet others have concentrated on part of a sector or sub-sector of the process industries, e.g. the pesticide industry in the chemical industry (Achilladelis *et al.*, 1987), ethylene manufacturing in the chemical industry (Hutcheson *et al.*, 1995), the plastics industry (Freeman *et al.*, 1963), the rubber industry in the chemical industry (Morris, 1989), oxygen steelmaking in the metallurgical industry (Lynn, 1982), smelting technology in the metallurgical industry (Särkikoski, 1999) and a single company like Du Pont in the chemical industry (Hounshell and Kenley Smith Jr, 1988)

The conclusions from many of the above examples of studies from different industries are that comparisons between industries from the process industries and across sectorial demarcation lines within the process industries have often been interesting and fruitful. Many studies have covered multisectorial samples from the process industries, and the results are often discussed from this standpoint with a view to applying the findings from a study to other process industries outside the group studied. In conclusion it has often been recognized in these studies that the process industries as a group do have several characteristics in common that make them worth researching and analysing as a whole.

Information about innovation in the process industries will be presented as references in the individual parts of this book. However, a review of the above-mentioned studies, and literature surveys, reveal that research into MOT can be complex. First of all it has been noted that different publications often use totally different sets of references. This is of course normal, but it is felt that in many cases older publications, presumably still of great importance, have been neglected and that important references have been missed. A review of the publications also revealed a number of uncertainties in the theories, models and proposed best working practices. Some weaknesses have been recognized not only by the present author but also by several distinguished researchers (Cobbenhagen *et al.*, 1990; Freeman, 1990; Leonard-Barton, 1992; Pisano, 1997; Tidd, 2001). The problems, according to this author's view, can briefly be summarized under the following headings:

Models are not always good pictures of the reality of innovation in industry

From an industry R&D professional point of view, many models do not seem to correspond well to an industrial reality. Sometimes the models may oversimplify a complex reality, making them unusable for industry. Sometimes the models or descriptions seem to be too industry-specific and lack a generic quality.

Models and theories sometimes point in different directions

Sometimes one gets the impression that the research results originate from different realities: what development actually looks like, what the R&D management would like it to look like, and what is an impossible average of several completely different R&D environments and development contexts (Skinner, 1992).

Irrelevant comparison of "apples and oranges"

Studying product and process development projects without recognizing their individual characteristics has created a lot of confusion when results from different studies are analyzed and compared. There is a lack of clear definitions of different kinds of innovation.

Failure to recognize and clearly describe the context in which development takes place

This is not an unusual phenomenon in emerging disciplines, but as a consequence of this, the results are often not comprehensible to or usable by either the academic researcher or industry professionals. This unfortunate situation makes many very interesting research results worthless to industry, raising barriers between academics and industry professionals and wasting valuable and important information. Lately, research on industrial research and development has to some extent been focused more on product development and how it is managed in manufacturing industry or service industries other than in the process industries, and sometimes with reference only to fast-moving consumer goods.

It should not be forgotten, however, that a large part of manufacturing industry, here called the process industries (the concept will be discussed in detail in Chapter 2), still operates in another industrial environment, using different production processes and producing different kinds of products. When product and process development in the process industries has been the subject of research in the past, the industry level of analysis has often been in focus (Malerba, 2004; Pavitt, 1984; Utterback and Abernathy, 1975), and with the historical perspective on the development of process technology (Utterback, 1994) or the influences of industrial innovation on society at large. Less research has been carried out on how to improve performance in process innovation and on the managerial tools for such improvements at a company level.

1.3 Some conceptual clarifications

This book uses material from different sources, and references are made to individual books and articles where different concepts have been used in different contexts. Because of that it has not been possible to be very stringent in the use of some concepts; in the following you will find guidelines to the practice selected for this book.

Firms, companies and corporations as industrial organizations

All the above concepts have been used as interchangeable names for industrial organizations. Further on, innovations within some functional areas of the firm like services, finance, etc. are not included — not because they are of less interest, but because their innovations are nevertheless something that is not closely related to process innovation.

Innovation and innovation-related activities

In the process industries, innovation activities are not confined to product development; a substantial proportion of corporate R&D must usually be devoted to process development, application development, applied research, technical support, etc. Not only that but there are a number of activities that cannot be defined as innovation in the process industries but are strongly related to successful process innovation. These have been designated "innovation-related" activities. In Chapter 3, a number of different kinds of innovation and innovation-related activities have been structured to illustrate the different kinds of R&D needs that exist in the process industries.

Different kinds of R&D: R&d, r&d, r&D, R&D

In this book, the concept of "innovation" in a firm is used in a sense similar to the more traditional concepts of "R&D" or "development". It is however well recognized that "innovation" should be seen from an overall company perspective for purposes of organization, resource allocation and management, and that R&D is only one aspect of the process of developing new ideas into marketable products, manufactured by more efficient production processes and sold by an eager market organization.

In the discussion of the newness of innovation, the term "incremental innovation" is used to refer to innovation of a lesser degree of newness compared with "radical innovation". The term "R&D" is often applied even to organizations that carry out research only or, as in most cases in the process industries, development only. In the headline the use of upper and lower case letters in "R&D" symbolizes the fact that in many R&D organizations there is a substantial difference in the combination of research and development that actually takes place.

Distinguishing between "product innovation" and "process innovation"

Most importantly, product innovation is considered as something distinct from process innovation; this will be discussed in depth in Chapter 3.

1.4 Some issues to reflect upon

This short introduction is simply intended to pave the way for the following parts and chapters of the book. At the end of each chapter, the contents will first be briefly summed up and then followed by some issues to reflect upon. This practice starts here with the presentation of the three following questions:

- *It has been stated that: "A good theory is the practitioner's best friend". Do you agree that a theoretical perspective can be a good "coat-hanger" for practical knowledge?*
- *Is innovation management recognized in your firm, and in what ways does your firm try to improve its knowledge in the area of management of innovation & technology?*
- *Has your firm ever tried to exchange information and learn good innovation behaviour from other firms in your sector of industry or from firms in other industrial sectors within or outside the process industries?*

Chapter 2

The Process Industries

"A class could therefore no longer be defined necessarily by the invariable presence of certain common attributes. Since the members of a class composed by sporadic resemblances were not assumed to be identical in any respect, it was no longer true that what was known of one member of a class was thereby known of the other members."

Rodney Needham (1975)

Writing a book about management of process innovation in the process industries requires, for multiple reasons, a careful analysis of the basic concept of "process industries". The reason is first of all that we must clearly distinguish not only between what kind of industries we are addressing, but also between the characteristic features of those industries that will influence innovation and innovation-related activities, even if all firms and industry sectors do not share all attributes.

The concept of process industries includes the two terms: "process" and "industries", and consequently one may expect the meaning of this concept to be related to a type of industry that uses some kind of process. A few decades ago there would have been no uncertainty. The designation process industries, as a subset of all manufacturing industry, was normally applied only to ones like the chemical and pulp & paper industries, etc. The introduction of new flow-oriented organizational structures, viewing activities within a company as transfunctional work processes (key business processes), has created a situation where all manufacturing industry is sometimes referred to as process industry. The term "process industries" will be preferred to "process industry" in this book to emphasize that this group of industries is not homogeneous but consists of a number of industry sectors more or less similar in individual characteristics. Needham's discussion, and the proposed grouping of industries belonging to the process industries in this book, also rely to a large extent on Wittgenstein's concept of family resemblance (Wittgenstein, 1992, p. 43).

However, if we now view the concept from an industry professional's point of view, we will probably come back to the old meaning, which refers to industries using a type of technological process. Even to people outside the industrial community, the term probably conjures up an image of an industry bristling with pipes and pumps and populated by men in hard hats who ought to be more concerned about the environment. Since this book deals with the subject area of managing process innovation in the process industries, it is important to give a clear and thorough industrial context for such an activity. This will be done via the formal exercise of defining this group of industries better. Disregarding the "hen and egg" situation, we will start with a general description of the process industries using a technological process perspective. On the basis of this description, a tentative comparison with other manufacturing industry will be made, two kinds of definitions will be proposed and the validity of the definitions will also be briefly discussed.

2.1 The production process and an intensional definition of the process industries

There are several differences between the process industries and other manufacturing industry (Dennis and Meredith, 2000; Taylor *et al.*, 1981), that often make not only general management but innovation management in particular rather dissimilar. As in all types of manufacturing industry, the production process in the process industries

can be described using a simplified input-output model of the process
and material flow:

- Type of incoming material: the feed to the production process
- Type of production process: the transformation process
- Type of outgoing material: finished products

Incoming material

Incoming material in the process industries is often called "raw materials". By lexicographic definition, this is "material before being processed or manufactured in a final form" (Webster, 1989). The term is often used in the process industries synonymously with "materials going into the plant", and does not necessarily mean untreated natural raw material. The expression can be used to distinguish from "components", which are manufactured parts and which often constitute the input to and sometimes also the output from other manufacturing industry.

It is practical to use "raw materials" in this broader sense and as a concept that distinguishes this group of materials from components. Another possible term could be "bulk material", an expression that recognizes that the material is not stored and transported in packages (Webster, 1989).

The production process

One characteristic feature, of significant importance to innovation in the process industries, is the production chain illustrated in Figure 2.1. The production chain, starting with raw materials, often includes intermediate deliveries of other finished products before final production and delivery of the end product to the customer. The production chain may include a number of large or small production plants, sometimes operated and owned by different companies, sometimes fully integrated in a long production chain within an industrial group or conglomerate.

Each production plant may include several separate production processes that are likewise connected in a chain structure. Each of

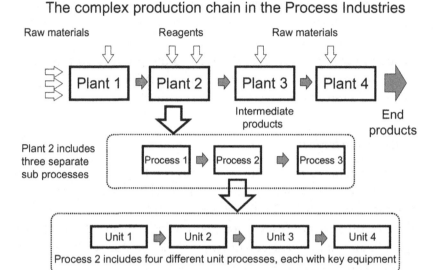

Figure 2.1   A schematic (Russian doll) model of the production process in the process industries. A production chain in the process industries is sometimes as complex as in other manufacturing industry (Lager, 2002).

those individual processes often also includes several unit operations, with certain key operational equipment. Production interfaces are sometimes created because of production conditions like availability of raw materials, transportation logistics, environmental considerations, nearness to end user, availability of cheap energy, etc.

Sometimes these interfaces arise from other causes, shaped by history or by mergers or other business events. The often long and complex chain structure of production units and interfaces described here can sometimes create artificial obstacles, preventing sound product and process development and disconnecting the total chain of customer demands on the products. Successful development of new products and new processes depends to a high degree on an understanding of this total chain structure (Tottie and Lager, 1995).

On the other hand the internal production chain with no formal supplier/customer interfaces may be too diffuse to clearly define the internal demands and needs for the intermediate products, with the result that the internal chain is sub-optimized because of internal discussions and disputes. The impact that innovation activities have on a company's total performance is likely to be associated with how much of this production chain is addressed in the development of new products and processes. A reduction of the number of process steps here may be a good indicator of what can be regarded as major process improvements (Utterback, 1994).

The integrated view of all plants and processes in a total production chain is growing more important in the context of supply-chain management and plant-wide process control, establishing a more efficient flow-oriented production process and information flow, aiming at improved overall company economy (Persson, 1997; Samskog *et al.*, 1995). The process chain structure, with its intermediate products and product deliveries, is also a chain of relationships between a chain of suppliers and customers that commonly occurs in the process industries. It is argued that this long supply chain is very common in the process industries, but can also be found in other types of manufacturing industry.

Outgoing material

There are certainly a number of similarities between the process industries and other types of manufacturing industry, but there are also a number of differences. One such difference is that firms in the process industries are often more sensitive to stoppages and interruptions in the process chain than other manufacturing industry, because of loss of production quality and long lead times for start up. One option is to classify the process industries according to the type of products that are produced. Utterback recognizes that there is a difference between "industries that produce assembled products" and "industries that produce non-assembled products" (1994, p. 103). Non-assembled products are also called homogeneous products, but it can be argued that an industry should not be defined solely in terms of what its products are not.

Divergent product flows

In the process industries the product flow in the supply chain often starts at upstream production facilities using *in-situ* raw materials of different character and quality. Such refined raw materials are then often delivered to midstream production plants where they are further processed and refined. Those processed products, and now often a considerable number of product qualities, are then further delivered to downstream production facilities where sometimes a very large number of products are made, with individual product specifications for different applications. In this book this situation has been designated a **"divergent product flow"** (Storm *et al.*, 2010).

The importance of supply-chain collaboration for improved innovation performance has been stressed in many studies (Cantista and Tylecote, 2008; Sahay, 2003; Soosay *et al.*, 2008), and such an internal and external supply chain is illustrated in Figure 2.2.

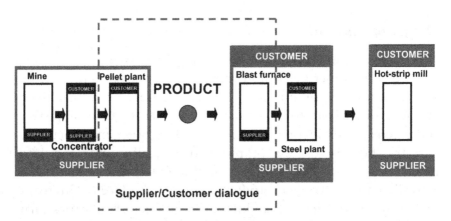

Figure 2.2   The complex supply chain for production of hot-rolled strip steel via the blast furnace route. Only a small part of the supply chain is illustrated in this figure. There are many supplier/customer interfaces, both internal and external (Tottie and Lager, 1995).

Serving mainly Business to Business customers but reaching for the end-user

Companies in the process industries, as intermediate midstream links in long supply chains, often lack direct contact with or even access to the end user. As such they normally have customers of Business-to-Business (B2B) type deploying large organizations of professional purchasers or traders, unlike other manufacturing industry, which mainly serves private consumers. The B2B kind of customer may be an excellent partner for the company and the company's development organization but can, in unfavourable circumstances, be an effective barrier to collaborative development and free information flow about future customer needs and product demands. Another characteristic of B2B customers for the process industries is that the sale of a product is not a single sale to one consumer, but often the sales of very large product volumes, sometimes delivered in bulk to a single customer.

An intensional definition of the process industries

Although there are surprisingly few clear definitions of the process industries, some can be cited. Chronéer (1999) discusses a few alternative possibilities ranging from an engineering perspective with pipes and pumps, and proposes a definition related more to development work. Another way of defining different classes of industry is to look at the ratio of the number of material varieties used to the number of products produced in the industrial production process. This type of classification is used in the "material conversion" classification system (Burbridge, 1995). An intensional definition gives the meaning of a term by specifying the properties required to come to the definition: the necessary and sufficient conditions for belonging to the set being defined (Foellesdal *et al.*, 1990). An intensional and more formal definition has been developed using the above brief outline of the production process in the process industries. The following intensional type of definition, which characterizes this industry in a descriptive manner, has been used in this book (Lager, 2000):

> "The Process Industries are a part of manufacturing industry using (raw) materials to manufacture non-assembled products in a production process where the (raw) materials are processed in a production plant where different unit operations often take place in a fluid form and the different processes are connected in a continuous flow."

This definition focuses on the production process and process technology characteristics that are probably foremost in the consciousness of industry professionals, but is probably also comprehensible to a wider public as well. Is "process industries" then the proper name for this group of industries, or should another name be selected that distinguishes it more clearly from other industries? Skinner uses the term "process industry" in distinction to "other manufacturing industry", which he calls "non-process industries" (Skinner, 1992). Bower and Keogh (1996) call industries from process sectors "process dominated industry". Another possible name could be "process based industry". Since the name "process industry" has been used for a long time and is still widely used by industry professionals, it is suggested this name should be kept but slightly modified to the plural, "process industries", indicating a cluster of more or less similar types of industry sectors.

If the sole criterion is that the material going into production is raw material and not components, the result might be that some types of industry with a long production chain starting with raw materials and ending up with assembled products would be unintentionally included. The concepts of unit operations, fluid form and continuous flow exclude industries that process solid raw materials but not in a process that would normally be associated with the process industries.

The criterion of non-assembled products excludes all types of industries that manufacture products by assembling components. If that were the only criterion however, some industries not normally considered as process industries would be included, such as mining, forestry, oil exploration, etc. Those industries are however excluded by the plant environment criterion. Bower and Keogh (1996) call those raw-material-producing industries "upstream" industries, as opposed to "downstream" industries. The latter are plant-based, concepts that also will be used in this book, supplemented with the concept "midstream" industries.

Some companies or conglomerates may have both upstream and downstream operations. As a consequence of this, whether or not they should be considered as a part of the process industries could then depend on what part of the organization is studied. The individual parts of the above definition can naturally be more or less satisfied insofar as the flow can be more or less continuous in the plant, the material can be more or less fluid and the number of individual unit operations few or many, etc. All individual parts or criteria in the definition may be satisfied, or only some of them, so the congruent validity is not as good as the discriminant validity. The main purpose of the definition, however, is not to decide which industries more or less belong to the process industries, but to serve as a guide for selecting industries that have operational or other similarities. Finishing with the discussion of the validity of the definition, we will continue to examine which industrial sectors, using this definition, could be included in the process industries.

2.2 An extensional definition of the process industries and evidence from the EU Scoreboard

While the previously presented intensional definition describes the characteristics of companies belonging to the process industries, it does not clearly state which industry sectors or companies should belong to this group; an extensional definition is more appropriate for this purpose. An extensional definition is the supplement to an intensional definition to the extent that it lists all items that fall under the definition. Sectors such as chemical, automotive and agricultural, etc., are made up of similar types of companies producing similar types of products and often using similar production processes. These sectors or groups of industries are also often selected for purposes of collecting and comparing statistical information. Industries within these sectors also often co-operate in informal structures or in national and international organizations and in many different areas, often including R&D.

Using the previously presented intensional definition of the process industries, a number of industrial sectors and industries have been selected from the total group of manufacturing industry presented in the *Statistical classification of economic activities in the European Community* (NACE, 2006).

In Table 2.1 a list of industries from the NACE system that are selected for inclusion in the category of process industries can be found. It can of course be argued that other groups should be included or that some industries should be excluded from this list, but in all cases this kind of classification can serve as a guideline. For a more detailed specification of the NACE codes, see App. A.

The 2008 EU Industrial R&D Investment Scoreboard, released in October 2008, presents information on 2000 companies from around the world that reported major investments in R&D (Guevara *et al.*, 2008). The set of companies it covers comprises the top 1000 R&D investors whose registered offices are in the EU and the top 1000 registered elsewhere, investing the largest sums in R&D in the last reporting year. The details of this database and the associated variables and terms are presented in App. B. The 2000 companies listed

Table 2.1   Selected Industries Belonging to the Process Industries. Process Industry Sectors (by NACE codes).

Oil and gas	06; 19
Chemical	20; 22
Utilities	35; 36; 37; 38
Pharmaceuticals	21 (not biotechnology)
Food	10
Steel	(24.1; 24.2; 24.3)
Mining & metal	05; 07; 24 (not 24.1; 24.2; 24.3; 24.5)
Mineral & material	08; 23
Beverage	11
Forest	17
Biotechnology	21

Table 2.2 The Process Industries as a Part of all Industry.

Variable	Industry grouping	Worldwide	Non European	European
Number of	All industry	1980	997	983
companies	Process industry	566	279	287
	Process industry (%)	29	28	29
Net sales	All industry	12,154	6,639	5,515
(€ billion)	Process industry	4,681	2,354	2,327
	Process industry (%)	39	35	42
Number of	All industry	43,972	23,674	20,298
employees	Process industry	10,154	5,112	5,042
(thousands)	Process industry (%)	23	22	25
R&D	All industry	379	253	126
investment	Process industry	108	70	38
(€ billion)	Process industry (%)	28	28	30

in this Scoreboard account for about 80 percent of worldwide business enterprise expenditure on R&D. Using the previously presented NACE codes for the process industries and reclassifying the companies included in the Scoreboard, a new list of firms belonging to the process industries has been compiled. In Table 2.2 the companies belonging to the process industries have been identified as one part of all industries. The figures are given on number of companies, net sales, number of employees and R&D investment.

Depending on which of the variables and geographical areas presented in Table 2.2 are selected, the process industries turn out to be a substantial part of all industry. To err a little on the conservative side, one can say that about 30 percent of the most R&D-intensive firms worldwide belong to the process industries.

2.3 The cyclic nature of the process industries and some industry sector specifics

Apart from the technological characterization of the process industries presented above, there are some other characteristics that are often particularly relevant to them in relation to innovation and process innovation, which will be briefly discussed in the following.

The cyclic nature of the process industries

Many sectors of the process industries usually experience some sort of cyclic nature of the market for their products (Rogers, 2005). This is sometimes referred to as "boom" or "bust" market behaviour (Morrison, 2005). Such behaviour occurs when the prices of products tend to go up and down in a pattern that is not so easy to predict. The last cycle of a boom character in the first decade of this new millennium was, to everybody's surprise, considerably extended over a longer time span than usual for the first time, and because of that was referred to as a "super cycle".

The mechanisms for this cyclic nature of the market for different kinds of products produced in different industry sectors will not be discussed further in this book and are left to the economists to explore and explain. The consequences of this behaviour of the market for products produced in firms that belong to the process industries are however strongly related to companies' innovation strategies. Products that are strongly dependent on the cyclic nature of the market are often those that can be characterized as "commodities", while products that sometimes are not so sensitive to those kinds of market fluctuations can be characterized as "functional products". Commodities surviving in the "bust" periods of the market crave a strong raw-material supply base and cost-efficient production technology (often delivered by efficient process innovation).

Commodities or functional products

In the original and simplified sense, commodities were things of value of uniform quality, produced in large quantities by many different producers. The items from each different producer were considered equivalent. It is the contract and this underlying standard that define the commodity, not any quality inherent in the product. Markets for trading commodities can be very efficient, and these markets will quickly respond to changes in supply and demand to find an equilibrium price and quantity. Examples of commodities or product denominated commodities include not only minerals, agricultural products and so on, but also so-called "commoditized" products like personal computers. Alternatively one could define commodities as products traded on commodity bourses like the London Metal Exchange (LME), Tokyo Commodity Exchange (TOCOM), Nymex, Tokyo Grain Exchange, etc. The following intensional definitions will be used in this book (Lager and Blanco, 2010):

> Commodity products are of uniform quality, with a low degree of differentiation which makes them more or less interchangeable. Prices are set on active markets that respond to changes in supply and demand. There are often many suppliers, and goods are easy to transport and store, often in bulk quantities. Customers are often business to business (B2B).

> Functional products have differentiated properties which mean that they are not normally easily interchangeable. Prices are set by suppliers on a cost-plus basis and not as a market price. Products are produced by a limited number of suppliers and they are not usually delivered in bulk quantities. Customers are often "end users" and sometimes consumers — business to consumers (B2C).

Innovation in the product downgrade-upgrade cycle

Firms in the process industries nowadays often try to avoid being simply commodity producers and strive to produce more functional products that offer more benefits to customers with higher profit margins. When products with improved functional properties are introduced on the market, they are usually imitated before long as competitors try to produce the same type of product with the same performance (Linn, 1984); prices then gradually decline, and the functional products degenerate into commodities, as illustrated in Figure 2.3.

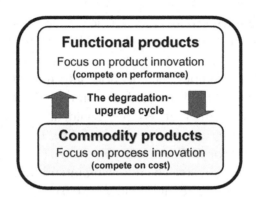

Figure 2.3   The arrows illustrate the degradation-upgrade cycle for functional and commodity products with different focus for development activities for different types of products (Lager, 2000).

Commodity products, i.e. products traded on the world market and mainly competing on price, usually need a focus on process development, while more functional products often benefit from product development (Cobbenhagen *et al.*, 1990). A cost-competitive commodity producer with the desire to produce more functional products will thus need abilities in both product and process innovation. The ability to be cost-competitive is related to some extent to a clear understanding of the dynamics of the total cost structure in the production process, but mainly to the ability to develop and introduce cost-efficient process technology in the production processes.

Product life cycles

The well-known concept of the "product life-cycle curve", which gives product profitability and sales volume as a function of product time on the market, is certainly also useful in the process industries. (For a good discussion of this product model, see for example the 1987 publication by Urban *et al.*)

The concept can be used on a product level and perhaps also on an aggregated product level such as "product line", but does not really make sense on a company level of analysis if the company is producing several different lines of products. The concept "product life cycle" can be applied to both commodities and functional products. The life cycles of many commodities do however often extend over a considerable space of time compared to functional products. The life cycle of commodity products may be extended indefinitely (see Figure 2.3), and sometimes functional products do not disappear at all, but only degenerate into commodity products with smaller profit margins.

Innovation characteristics from the Scoreboard

Table 2.3 shows the companies in the Scoreboard separated into 11 sectors using the classification for the Scoreboard and NACE codes, similar to the format for Table 2.2. Monetary and percentage figures for all sectors are grouped in ascending order by their figures on net sales.

2.4 Summing-up and some issues to reflect upon

This chapter started with a thorough discussion of the concept of process industries, with an intensional definition and presentation of some technological features that characterize this cluster of industry sectors. The object was, however, not only to start with a clear concept and definition for this book, but to make a presentation of important technological attributes that may affect process innovation in this group of industries. The material conversion perspective was selected because the influence of parameters associated with the material flow to a large extent give, or should give, one important contextual framework for not only process innovation and process innovation strategies, but firm innovation in general (Storm *et al.*, 2010).

Using this intensional definition, a number of industry sectors or sub-sectors were identified as belonging to this cluster. Starting with a classification by their statistical codes and information from the Scoreboard, it was discovered that an estimated 30 percent of the 2,000 most R&D-intensive firms in industry worldwide belong to the process industries cluster. Using the concept of family resemblance,

Table 2.3 Industry Sectors in the Process Industries by Net Sales, Number of Companies, R&D Investments and Number of Employees. Figures from the Scoreboard (Guevara *et al.*, 2008).

Sector	Net sales		Number of companies		R&D investment		Number of employees	
	€ billion	%	Numbers	%	€ billion	%	Thousand	%
Biotechnology	38	0.8	121	21.4	9.2	8.6	122	1.1
Forest	81	1.7	10	1.8	0.4	0.4	255	2.4
Beverage	90	1.9	7	1.2	0.5	0.5	442	4.1
Mineral & material	129	2.7	23	4.1	1.2	1.1	474	4.4
Mining & metal	176	3.8	17	3.0	1.3	1.2	620	5.7
Steel	292	6.2	21	3.7	2.0	1.9	857	7.9
Food	326	7.0	52	9.2	4.4	4.0	1172	10.8
Pharmaceutical	427	9.1	120	21.2	60.5	56.1	1430	13.2
Utilities	633	13.5	44	7.8	3.3	3.1	1496	13.6
Chemical	672	14.4	125	22.1	19.0	17.7	1882	17.4
Oil & gas	1818	38.8	26	4.6	5.8	5.4	2115	19.5
Total	4681	100.0	566	100.0	107.7	100.0	10.835	100.0

it is argued that industries in this cluster may not share all similar characteristics, but so many that they can be identified as a family of industries that can be discussed and researched as an interesting group.

Finally, some other process industry characteristics are taken up and the market perspective on products is introduced. The product downgrade-upgrade cycle is discussed, and the life-cycle perspective on commodities and functional products is presented. The importance of process innovation to a cost-competitive commodity producer is recognized.

- *Using the previous classification and definition, which industry sector of the process industries does your firm belong to?*
- *How homogeneous is that sector in your firm's competitive perspective?*
- *What kind of products or product groups (commodities or functional products) does your firm manufacture?*
- *Who are your firm's main competitors and how much do they spend on innovation?*
- *Studying the figures in Table 2.3 on R&D investments. How does your firm compare to the sectorial average?*
- *How important is innovation in your firm and how much innovation is enough for your firm to stay competitive in the future over a business cycle?*
- *Using the presented schematic technological description of the production process of the process industries, which are the main characteristics from this material conversion perspective that affect your process innovation activities the most?*

Chapter 3

Innovation in the Process Industries

"When ambiguity, vagueness, etc. are a hindrance rather than a help to communication and to motivating and discussing different points of view, we need to clarify what we mean. We can do this in several ways. We can for example use other words and expressions which are more precise, or with which the other party is more familiar. However, we can also continue to use the same words, but explain what we mean by them. Such an explanation is often called a definition, from the Latin definitio (delimitation)."

Dagfinn Föllesdal *et al.* (1990)

In Drucker's plea for higher productivity for knowledge workers (1998, p. 147) he recognizes that the difference between the old activities of "making and moving things", and the activities of today's knowledge workers, is that the tasks used to be taken for granted. For today's knowledge workers, the first step on the road to productivity improvements is thus to define the task. This may seem trivial, but that is certainly a misconception.

This chapter therefore starts with a through discussion and introduction of different innovation and innovation-related concepts and activities in the process industries, and also refers to the problems described in the introduction to Chapter 1. Not only are the classical innovation activities like product and process innovation defined, but new areas of innovation are introduced, in particular development of raw material supplies. Starting with a discussion of such "backward integration" and raw material development in the supply chain, activities of "forward integration" are also introduced. Such "application development" is defined as the development of the customer's use of the firm's products; an important area that so far has received far too little attention in academic publications.

Apart from these areas of innovation, some linked activities, here called innovation-related areas, are also presented and defined. The background is that they are of considerable importance to all firms in the process industries, and to process innovation in particular. The importance and the objective of defining innovation concepts more clearly is to better understand the different skills and resources they require, but certainly not to keep them operationally more apart! Improved integration between these different activities is highlighted in a following section, and will be further discussed in Chapter 4.

Innovation intensities in different sectors of the process industries are presented in the final section. Statistical evidence from the European Scoreboard provides information from 2000 large firms worldwide, while the results from a European research project give evidence, particularly from the Swedish process industries.

3.1 An improved typology for innovation and innovation-related activities in the process industries

The work in an R&D organization is often of a wide and disparate nature, often difficult to fully grasp in simple statements and definitions. Figure 3.1 shows that some activities cannot be regarded as development at all, but they are nevertheless often included in the area of responsibility of the R&D department and are consequently often included in the annual R&D budget. Later we will designate all activities that are usually performed in the R&D organization in the process industries as innovation (development) or innovation-related (development-related) activities.

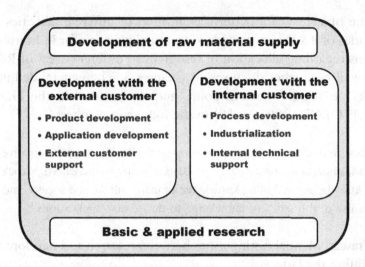

Figure 3.1 A proposed classification of different types of innovation and innovation-related activities often performed in the R&D organization in the process industries. Observe that some activities cannot be classed as R&D according to strict OECD definitions (OECD, 2002).

Figure 3.1 lists various kinds of activities often carried out in the R&D organization in the process industries and tentatively groups them into four clusters (Lager, 2008):

- Development of raw material supply.
- Development with the internal customer.
- Development with the external customer.
- Basic research and applied research.

Much confusion is caused both in company R&D departments and in academic studies of company R&D because comparisons are made between completely different categories and types of R&D activities. It is consequently of common interest and importance to make those categories clear, understandable, and also more useful to industry professionals.

One objective of a better classification of different activities and categories of R&D is to provide a structural platform for in-house discussions regarding allocation of resources to different areas of R&D, which will be further discussed in Chapter 5. The general definition of R&D is a good starting point, and is expressed in the Frascati Manual (OECD, 2002, p. 30) in the following manner:

> "Research and experimental development (R&D) comprise creative work undertaken on a systematic basis in order to increase the stock of knowledge, including knowledge of man, culture and society, and the use of this stock of knowledge to devise new applications."

The Frascati Manual distinguishes between research and development by defining the latter:

> "Experimental development is systematic work, drawing on existing knowledge gained from research and/or practical experience, which is directed to producing new materials, products or devices, to installing new processes, systems or services, or to improving substantially those already produced or installed. R&D covers both formal R&D in R&D units and informal or occasional R&D in other units."

The advantage of using the OECD definitions is that consistency is ensured, company-specific definitions can be avoided and other loosely defined concepts like "curiosity driven research" are made redundant.

Basic and applied research

Basic and applied research are activities that are now often left to external organizations outside the company's R&D organization. The Frascati Manual (OECD, 2002, p. 30) states:

> "Basic research is experimental or theoretical work undertaken primarily to acquire new knowledge of the underlying foundation of phenomena and observable facts, without any particular application or use in view. Applied research is also original investigation undertaken

to acquire new knowledge. It is, however, directed primarily towards a specific practical aim or objective."

Development with the external customer

Product development is often the main development activity undertaken jointly with the external customer.

Product development (product innovation)

The Oslo Manual (OECD, 2005, p. 48) gives a general definition of product development:

> "Product innovation is the introduction of a good or service that is new or significantly improved with respect to its characteristics or intended uses. This includes significant improvements in technical specifications, components and materials, incorporated software, user-friendliness or other functional characteristics."

In this definition it is important to notice the need for "significantly improved" products. Product development is normally carried out in collaboration with the customer's production organization, but intimate collaboration with the customer's product development teams will be desirable in the future.

Product and process development is sometimes looked upon as the same activity in the process industries. It can be argued that it is not necessary to distinguish the work processes for "product development" and "process development" because product development in the process industries also partly takes the form of development work in a laboratory. The strongest argument against that point of view is that both those work processes start with different customers and end up with different customers, a fact that will be further discussed in Chapters 8 and 9. The following slightly modified complementary definitions underline this idea (Lager, 2002):

> "Product development is defined as development driven by a desire to improve the properties and performance of finished products,

even if the nature of the practical development work is sometimes development of process technology in a laboratory. Objectives for product development can be improving product properties, improving product quality (uniformity of composition), environment-friendly products, etc. Product development is then also the customer for the development of necessary process technology to produce the desired new or improved products."

Another, often very important development area in the process industries, is development with the external customer: that is, application development.

Application development

Application development can sometimes be considered as "the low-hanging fruit" of development work because it does not involve costly investments in company process technology or the development of new products. The failure to identify and distinguish application development from product development can however be very serious.

The abilities the R&D organization needs for application development are thus quite different from product development. The expert knowledge that is needed is not about one's own company's processes and product development, but those of the customer. It is first of all important to note that not all application areas may require a collaborative approach with the customers at the level of joint collaborative application development.

The search for new applications and new customers for existing products is thus the prime responsibility of the marketing organization, and is a different kind of activity that precedes the actual application development. A definition of "application development" that will be used in this book is:

Application development is primarily the significant development of the customer's use of the supplying company's own products. The development work is an optimization of the customer's production system in the use of the company's products. It may improve the customer's process and/or products. Application development thus implies an involvement in the customer's process and product development.

External customer support sometimes also named "market support" or "technical services"

External customer support comprises all kinds of activities that aim to help the customer to use the company's products in an efficient manner. There are however no development activities associated with this kind of service, and services are restricted to existing products. Activities in this area are sometimes designated as part of the "meta-product" and as such are activities that can differentiate commodities or commodity-like products from those of competitors. There is probably no simple way to define "customer support", but each company should include whatever selected spectrum of supporting activities it finds useful. A definition that will be used in this book is:

> Customer support comprises activities that are carried out primarily to help the customer in his/her use of the supplying company's existing products. Activities that often are included are training of customer representatives and seminars on the customer's premises. Customer complaints and other comments on delivered products are often, via market channels, investigated in this organization. No development activities are carried out on the firm's products or in the customer's processes.

Customer support activities are normally carried out in collaboration with the customer's production or purchasing organizations.

Development with the internal customer

Process development is, or should normally be, the main development activity carried out together with the firm's internal customer.

Process development

Process development can be defined according to the Oslo Manual (OECD, 2005, p. 49):

> "Process development (process innovation) is the implementation of new or significantly improved production or delivery methods. This includes significant changes in techniques, equipment and/or software."

Process development is thus much more of an in-house affair compared to product development. It is obvious from the above definitions that product development differs from process development, and it is concluded in the Oslo Manual "that with respect to goods, the distinction is clear". However, in the process industries it is sometimes found that the borderline between product development and process development ought to be better delineated and consequently slightly differently defined (Lager, 2000). The development of necessary new or improved process technology to produce new or improved products should therefore be included in the product development concept and its work process. This is in order to facilitate development integration and shorten time-to-market. In this case the customer for process development is also an internal customer: product development and the product development project. The following slightly modified complementary definitions underline these ideas (Lager, 2002)

> "Process development is defined as development mainly driven by internal production objectives. Such objectives may be reduction of production costs, higher production yields, improvement of production intensities, environment-friendly production, etc. In many sectors of the Process Industries, process development is mainly prompted by the needs of production (internal customer). Another internal customer to process development is the company's own product development."

The most important development activity conducted with the internal customer is process development, and development with equipment suppliers is therefore included in the process innovation concept. A noteworthy observation on the collaborative development of process technology/equipment is that it may be called either product development or process development depending on the viewpoints of the parties concerned (Rosenberg, 1982, p. 4). From the equipment supplier's perspective this kind of development is often discussed in terms of entering into a "product development project", whereas from the process firm's perspective it is typically discussed in terms of entering a "process development project". It may, however, be advisable for both the equipment supplier and the process firm to speak in terms of developing both a "product concept" and a "process concept". That is, for the process firm its product development is prompted by the needs of its customers for improved process technology, which as a consequence may prompt a need for the development of new process technology.

A similar situation typically occurs for the equipment supplier when the development of a new process technology for the customer (process firm) prompts the need for the development of a new product (new equipment). The customer's use of a process firm's already developed products is usually called "application development" in the process industries. In a similar vein, the use of the equipment supplier's product in the customer's process may thus also, when the product is further marketed to other customers, be regarded as application development and as improvement of the customer's use of the product. The consequence for the process firm of using this "mental map" is that it focuses development activities more on improvements of the customer's process than on the development of the actual product. The consequence for the equipment manufacturer may be that it focuses development activities more firmly on improvement of the customer's production gains than on the actual development of the equipment.

Two other innovation-related activities are industrialization and internal technical support.

Industrialization

The term "Industrialization" usually covers all kinds of design activities, installation of new process technology, erection and start-up of new plants. Some companies, and smaller companies in particular, choose to outsource all this work and thus do not have these types of resources in-house. This is however a drawback, because implementation of new technology has proven to be a weak area in process innovation; see Chapters 8 and 9 about process innovation work processes. Instead of defining industrialization intensionally, we give an extensional definition by enumerating activities often carried out in such an organization:

- Management of investment projects.
- Design of new plants.
- Erection of new plants and production units.
- Reconstruction of existing plants.
- Introduction of new products or processes in existing production plants.
- Introduction of production technology for new or improved products.
- Introduction of production technology for new or improved processes.
- Startup of new process technology or production plants.

The latter area, startup capabilities, will be more extensively treated in Chap. 12 because this is the final phase of all kinds of process innovation work processes.

Internal technical support

Internal technical support and industrialization are two activities which often go hand in hand. In this book internal technical support is defined as:

> "The improved use of well-tried firm process technology as an activity of a firm's day-to-day management of technology. When any process development efforts are necessary to reach desired targets and outcomes, the activity should be reclassified as process development."

Internal technical support may include a diverse number of activities and services that each firm has discretion to decide upon. It is thus important to define it carefully and to separate this area from process development in allocating resources. In the process industries, internal technical support often tends to absorb much more of available company resources than its strategic importance merits. Apart from general technical support in the form of contracted investigations and assignments, trouble-shooting in production plants and training are often included. Often other general services like information search and patent applications also fall within such an organization's remit.

Development of the raw material supply

The development of raw material supplies is an area of utmost importance to firms in many sectors of the process industries. For companies producing upstream commodities or those that are vertically integrated starting with raw material *in situ*, the development of a captive supply of raw materials is a most important area for sustainable long-term survival. For downstream companies producing more functional products, purchased raw materials may also influence the overall profitability of operations, and for some sectors the quality of finished products. The collaborative development of raw materials together with raw material suppliers is thus an activity of significant importance strongly related to process innovation. Lacking other more generally accepted definitions, the following working definitions will be used:

Development of captive raw material supplies
(*supplies owned by the firm*)

The development of captive raw material supplies is either the search for new raw material resources or the development of improved process technology for processing of such materials in order to extend the raw material base.

Development with raw material suppliers

Development with external raw material suppliers, with a view to improving the firm's "processability" of such materials. Raw material suppliers can thus call this activity "application development".

3.2 A need for a more dynamic interaction between different categories of innovation

In the process industries, including many sectors and a large part of all manufacturing industry such as minerals and metals, pulp and paper, food, chemicals and petrochemicals, etc., a substantial part of product development is not radical new product development but the incremental development and refinement of existing products. Thus in a typical company there are few radically new product developments during any given year, but many product improvements. For example, a product like a sheet of A4 office paper may seem like a rather simple product, but in fact it has about 40–50 measurable product properties available for improvement (Liiri-Brodén, 2005). This is further illustrated in Fig. 3.2 as the never-ending interaction between customers and suppliers. The embryos of future desired properties are conceived in such collaboration, and the development cycle never ends.

It is important to collaborate with the customer in seeing how the firm's own products could be used efficiently in the customer's production process, a procedure which is called "application development". Further collaboration must then take place to allow an understanding of how to translate today's and tomorrow's demands and expectations into measurable product properties, which we call the Voice of the Customer (VOC).

Next comes an appreciation of how this information can be translated into new and improved product concepts — understanding how functionalities are created. Further on it is necessary to be aware of how such new product functionalities and concepts can be produced within necessary product specifications and with a constant quality in cost-efficient production systems.

Knowledge of the customer's products and production technology

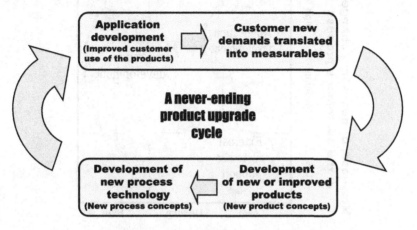

Figure 3.2   The never-ending product development cycle in the process industries. The interplay between product innovation, process innovation and application development is illustrated as a firm knowledge building process.

A process development project can thus give opportunities for product development, just as the development of new products can sometimes be combined with process development and cost reduction in the production process. A simplified test of which kind of development a selected project belongs to is to ask: If the project is successful, is the external customer or the internal customer going to be happy? If both are happy, the project is an example of the very special and important case of integrated product and process development, a highly desirable position in the world of innovation in the process industries.

As such, product development should use the skills of the application developers in a product launch and introduction to the customer. Nevertheless, the project should be classified in the company's internal work as a product development project, since in this case the application development team is to be considered as an internal resource to the product development team. Note the similarity to product development and process development. There is also often a

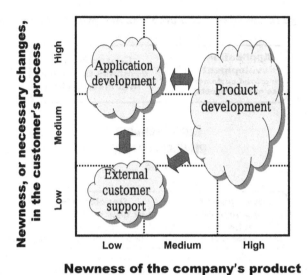

Newness of the company's product

Figure 3.3 Categories of joint development and development-related activities with external customers.

dynamic interaction between development projects. A project may start as a process development project but may finish up as a product development project, or vice versa.

The matrix in Fig. 3.3 was created as an aid to clearer classification of collaborative activities with the customer, using the dimensions "newness of the company's product" and "newness, or necessary changes, in the customer's process".

A distinction between different kinds of development activities is obtained by positioning them in this newness matrix. There are of course "grey zones" or even overlaps. The dynamic interaction between the different areas is symbolized by the arrows. The matrix was reduced to three areas:

- *External customer support*
  External customer support, using the previous definition, is about existing company products that are used by the customer in his present process (bottom-left corner in matrix).

- *Product development*
  Product development, from the previous definition, is always
  about a medium-to-high newness of the product. It is normal that
  there is a need to tune-in the customer's process when starting to
  use a new product, but in rare cases a new product will not need
  adjustments of the customer's process.
- *Application development*
  Application development, using the previous definition, is about
  adapting the customer's process to existing products. During
  application development it is not uncommon for a product to
  experience what one would like to call a "specification creep",
  like "scope creep" in project management, and that one or
  two specifications of a product are changed into a variant of the
  product.

3.3 Innovation intensities in individual industry sectors of the process industries

Innovation intensity is normally defined as a company's total annual
spending on R&D divided by its total annual turnover, expressed as a
percentage. This is normally one of the few measurable characteristics
of innovation presented in a company's annual report. The figure gives
the necessary resources allocated to innovation, or in plain words,
should give an overall indication of a firm's overall need for innovation.

Evidence from the EU industrial R&D investment scoreboard

The figures presented in the following Figs. 3.4 and 3.5 are taken from
the Scoreboard (Guevara *et al.*, 2008), of which further details can be
found in App. B and in Chapter 2. Figure 3.4 shows the R&D intensities of the largest R&D spending companies (in absolute figures) in the
process industries worldwide from this sample. The categorization of the
R&D intensities has been selected as low, medium-low, medium-high,
high and extremely high.

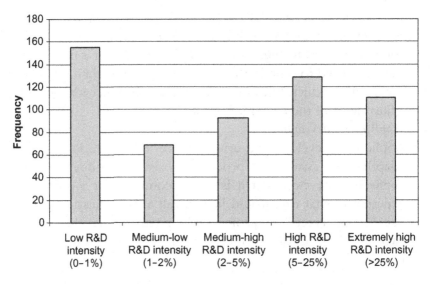

Figure 3.4   A categorization of R&D intensities for firms in the process industries in this sample from the Scoreboard. Figures from the Scoreboard (Guevara *et al.*, 2008).

	R&D/Net sales ratio 2007 %							
Type of process industry	Mean	Std	Q1	Median	Q3	Maximum	Skewness	N
Oil & Gas	0.4	0.3	0.2	0.3	0.6	1.2	1.4	26
Beverage	0.6	0.7	0.1	0.2	0.8	2.0	1.8	7
Forest	0.7	0.6	0.3	0.5	0.7	2.1	2.4	10
Utilities	0.9	1.3	0.3	0.6	0.8	7.4	3.7	43
Steel	0.9	0.6	0.5	0.8	1.2	2.6	1.4	21
Mining & metal	1.1	1.6	0.3	0.8	1.2	6.6	3.2	16
Mineral & material	1.2	1.0	0.5	0.9	1.9	3.6	1.4	23
Food	1.6	2.4	0.6	0.9	1.7	13.7	3.9	52
Chemical	3.0	2.3	1.5	2.5	3.6	19.0	3.3	125
Pharmaceutical	12.2	6.3	7.0	12.7	17.8	25.0	-0.1	91
Biotechnology	13.3	5.9	7.6	11.9	19.6	24.7	0.4	32
Sum								446

Figure 3.5   Innovation intensities in firms in different sectors of the process industries.

The large number of firms from different process industries sectors in the first column represents firms with a large turnover but a low R&D intensity. In Fig. 3.5, on the other hand, the firms have been separated into different sectors of the process industries.

The sectors are listed by the mean values on R&D intensities in ascending order. It is apparent that the sector spending the least on innovation, in innovation intensity figures, is oil and gas. However, the very large figures for company turnovers in this sector mean a sizeable amount of R&D in absolute figures. Nevertheless, the low percentage gives food for thought. The three other industry sectors that fall in the low intensity group are beverage, forest and utilities. The food and chemical industries show high figures for standard deviations and also for skewness. The very high figures for R&D intensity in the pharmaceutical industry and in biotechnology firms have long been widely recognized.

Evidence from the European research project

The European Research Project (ERP) was a large research project on process innovation in the European process industry; it included a large survey of R&D managers in different sectors. The project and methodology are presented in detail in App. C.

The R&D intensities were calculated using the respondents' figures for total company turnover and the figures for company R&D expenditure. The results for different categories of process industry are presented in Fig. 3.6. It can first of all be noted that most of the values are below 4 percent for all companies included in this study. All medians for all categories of industry are below 3 percent and the overall median figure for the total sample is 1.5 percent (arithmetical averages are also given in parentheses).

The median figure is highest for other types of process industry, 3.0 percent (4.6 percent); the second highest is chemicals, 2.2 percent (3.0 percent), followed by mining and minerals, 1.6 percent (1.6 percent), basic metals, 1.0 percent (1.3 percent), food and beverage, 0.8 percent (1.2 percent) and lowest for pulp and paper, 0.7 percent (1.2 percent). The overall impression is that there is a considerable spread in the values within each sector of the process industries. For example, the mining and mineral industry median for R&D intensity in this study is 1.6 percent, but the lowest figure is 0.2 percent and the highest is 3.6 percent.

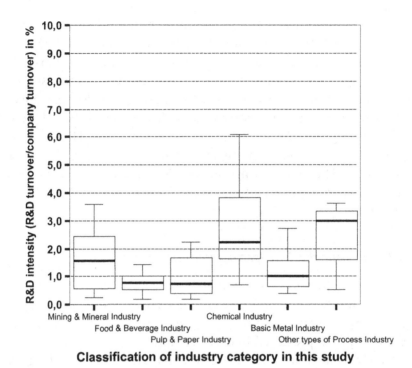

Figure 3.6 R&D intensities for different categories of process industry presented as box plots. The bold horizontal line in each box shows the median for the category. The box represents the interquartile range (50 percent of all values). The whiskers show the largest and smallest observed values.

The average R&D intensities for different categories of industry are interesting for companies to compare with, but the dispersion of R&D intensity within each category is even more so. The explanation for the wide dispersion of R&D intensity within each category may be that the breakdown into only six different types of process industry is too coarse. Another explanation is that such large differences actually exist even within finer categories (sub-sectors). The individual cases in this study appear to support the latter explanation.

There may probably also be other factors influencing R&D intensity within the same category, such as different corporate and R&D strategies, as well as company-specific fluctuations in R&D expenditures which may, but not necessarily must, have a cyclic pattern. It must always be remembered that R&D intensity is heavily influenced by fluctuations in a company's annual turnover. The R&D intensity can also be illusory, since a large company with low R&D intensity but with a high turnover still spends a sizeable amount on R&D in absolute figures, which will be reflected in its R&D organization and its innovative capacity. On the other hand, a small company with a high R&D intensity but a low turnover cannot undertake large development projects. This suggests that R&D expenditures should also be considered in terms of absolute figures.

Because the public statistics use a different classification system, and the sample is taken from a different population, a strict comparison is not possible with public statistics. Comparing the results from the study, one can only say that the figures agree fairly well with the figures from the Scoreboard for some industries, but differ substantially for others. One may ask whether figures from public annual statistics for different sectors of industry could, or should, be used as a guideline for individual companies. The results from this study show that a company must be careful when comparing its R&D intensity with published statistics for its category, because there seems to be a large spread of figures within each sector. A comparison of company R&D intensity with selected competitors would appear to be more relevant. Each company should carefully consider its past and future figures for R&D intensity (Cooper, 1988), but it must also be remembered that the amount of money spent on R&D is not the only important parameter. How well the money is used is another matter which will be discussed in Part 5. Nevertheless, it is felt that R&D intensity is an important yardstick for companies, provided that it is used cautiously, is monitored over an extended period of time, and is compared with selected competitors.

3.4 Summing-up and some issues to reflect upon

The importance of a corporate consensus on the use of innovation concepts and definitions is stressed in order to avoid comparisons between apples and oranges in innovation. An improved typology of innovation activities in the process industries is presented and will also form the backbone for all further discussions in the following chapters of this book including resource allocation and strategy, organizational structures and external collaboration. Furthermore, the development of success factors and measurables for process innovation will rely heavily on those well-defined concepts. A better classification of innovation activities is however not intended to keep those activities apart, but to stimulate better integration, and thus the dynamic interaction between different innovation activities is advocated.

It is hoped that the figures that are presented for innovation intensities in different sectors of the process industries will give food for thought. It is advisable to keep track of company innovation intensities over an extended period and also to compare only with competitors in the same sector or sub-sector of industry.

The dependency of the R&D intensities on changes in a company's turnover and the cyclic nature of some company innovation activities make it necessary to compare R&D resource allocations in absolute figures as well.

- *Reviewing the different innovation and innovation-related activities presented in this chapter, are all of them well recognized in your firm, and if not do you think that they should be?*
- *To what group of firms does yours belong regarding innovation intensity: low, medium-low, medium-high, high or extremely high?*
- *Is this the group that you believe your firm should also belong to in the future?*
- *Do all plants, companies, business areas and divisions in your corporation use the same concepts and definitions for innovation and innovation-related activities?*
- *Have the necessary different capabilities and correspondingly good innovation behaviour for the various activities been identified?*
- *The interaction between different innovation and innovation-related activities is stressed in this presentation. How can information, communication and collaboration between different innovation activities be improved in your firm?*

References

Achilladelis, B (1975). History of UOP: From petroleum refining to petro-chemicals. *Chemistry and Industry.*

Achilladelis, B, Schwarzkopf, A and Cines, M (1987). A study of innovation in the pesticide industry: Analysis of the innovation record of an industrial sector. *Research Policy,* 16, 175–212.

Achilladelis, B, Schwarzkopf, A and Cines, M (1990). The dynamics of technological innovation: The case of the chemical industry. *Research Policy,* 19, 1–34.

Aftalon, F (2001). *A History of the International Chemical Industry: From the Early Days to 2000.* Philadelphia: Chemical Heritage Press.

Allen, JA (1967). *Studies in Innovation in The Steel and Chemical Industries.* Manchester: Manchester University Press.

Argyris, C (2002). Double-loop learning, teaching, and research. *Academy of Management Learning and Education,* 1, 206–218.

Birkinshaw, J, Hamel, G and Mol, MJ (2008). Management innovation. *Academy of Management Review,* 33, 825–845.

Booz Allen and Hamilton. (1982). New Product Management of the 1980s.

Bower, DJ and Keogh, W (1996). Changing patterns of innovation in a process-dominated industry. *International Journal of Technology Management,* 12, 209–220.

Burbridge, JL (1995). *The Material Conversion Classification.* London: Chapman & Hall.

Cantista, I and Tylecote, A (2008). Industrial innovation, corporate governance and supplier-customer relationships. *Journal of Manufacturing Technology,* 19, 576–590.

Chakrabarti, AK (1988). Trends in innovation and productivity: The case of chemical and textile industries in the U.S. *R&D Management,* 18, 131–140.

Chronéer, D (1999). Customer-oriented trend in steel and pulp/paper industries? *Division of Business Administration and Social Sciences. Division of Industrial Organization.* Luleå, Luleå University of Technology.

Cobbenhagen, J, den hertog, F and Philips, G (1990). Management of Innovation in the Processing Industry: A Theoretical Framework. In C Freeman & L Soete (eds.), pp. 55–73. London: Printer Publishers.

Cooper, RG (1988). *Winning at New Products.* London: Kogan Page.

De Margerie, V (2009). *Strategy and Technology: Towards a Technology Based Sustainable Competitive Advantage.* Paris: L'Harmattan.

Dennis, D and Meredith, J (2000). An empirical analysis of process industry transformation systems. *Management Science*, 46, 1085–1099.

Drucker, P (1998). *On the Profession of Management*. Boston: Harvard Business School Press.

Enos, LJ (1962). *Petroleum Progress and Profits*. Cambridge, MA: MIT Press.

Foellesdal, D, Walloe, L. and Elster, J (1990). *Argumentationsteori Språk Och Vetenskapsfilosofi*. Oslo: Universitetsförlaget.

Freeman, C (1990). Technical Innovation in the world chemical industry and changes of techno-economic paradigm. In *New Explorations in the Economics of Technological Change*, C Freeman & L Soete (eds.), pp. 74–91. London: Pinter Publishers.

Freeman, C and Soete, L (1974). *The Economics of Industrial Innovation*. London: Continuum.

Freeman, C, Young, A and Fuller, J (1963). The plastics industry: a comparative study of research and innovation. *National Institute Economic Review*, 26, 22–60.

Guevara, HH, Tübke, A and Brandsma, A (2008). The 2008 EU Industrial R&D Investment Scoreboard. Luxembourg: Office for Official Publications of the European Communities. http://iri.jrc.ec.europa.eu (accessed 2008).

Hounshell, DA and Kenley Smith Jr, J (1988). *Science and Corporate Strategy. Du Pont R&D, 1902–1980*. Cambridge: Cambridge University Press.

Hutcheson, P, Pearson, AW and Ball, DF (1995). Innovation in process plant: a case study of ethylene. *Journal of Product Innovation Management*, 12, 415–430.

Kaplan, RS (1998). Innovation action research: creating new management theory and practice. *Journal of Management Accounting Research*, 10, 89–118.

Kline, SJ (1985). Innovation is not a linear process. *Research Management*, July–August, 36–45.

Lager, T (2000). A new conceptual model for the development of process technology in process industry. *International Journal of Innovation Management*, 4, 319–346.

Lager, T (2001). Success factors and new conceptual models for the development of process technology in process industry. Department of Business Administration and Social Science. Division of Industrial Organization. Luleå, Luleå University of Technology.

Lager, T (2002). Product and process development intensity in process industry: A conceptual and empirical analysis of the allocation of company resources for the development of process technology. *International Journal of Innovation Management*, 4, 105–130.

Lager, T (2008). Using multiple progression QFD for roadmapping product and process related R&D in the process industries. *14th International Symposium on Quality Function Deployment*. Beijing; China.

Langrish, J, Gibbons, M, Evans, WG and Jevons, FR (1972). *Wealth from Knowledge. A Study of Innovation in Industry*. London: Macmillan.

Leonard-Barton, D (1992). The factory as a learning laboratory. *Sloan Management Review*, 34, 23–38.

Lewin, K (1946). Action research and minority problems. *Journal of Social Issues*, 2, 34–46.

Liiri-Brodén, E (2005). HOLMEN.

Linn, RA (1984). Product development in the chemical industry: a description of a maturing business. *Journal of Product Innovation Management*, 2, 116–128.

Lynn, LH (1982). *How Japan Innovates — A Comparison with the U.S. in the Case of Oxygen Steelmaking*. Boulder, CO: Westview Press.

Malerba, F (ed.) (2004). *Sectoral Systems of Innovations. Concepts, Issues and Analyses of Six Major Sectors in Europe*. Cambridge University Press.

Miles, MB and Huberman, AM (1994). *Qualitative Data Analysis; An Expanded Sourcebook*, Thousand Oaks, CA: Sage Publications.

Morris, PJT (1989). *The American Synthetic Rubber Research Program*. Philadelphia: University of Pennsylvania Press.

Morrison, K (2005). Boom sending ripples through world markets (November 22). *The Financial Times*.

NACE (2006). *Statistical Classification Of Economic Activities In The European Community Rev 2*. Luxembourg: Office for Official Publications of the European Communities.

Needham, R (1975). Polythetic classification: convergence and consequences. *Man (N.S.)*, 10, 349–369.

Norling, PM (1997). Structuring and managing R&D work processes — Why bother? *CHEMTECH*, October, 12–16.

OECD (2002). *Frascati Manual: Proposed Standard Practice for Surveys on Research and Experimental Development*. OECD Publishing.

OECD (2005). *Oslo Manual: Guidelines for Collecting and Interpreting Innovation Data*. OECD Publishing.

Pavitt, K (1984). Sectoral patterns of technical change: Towards a taxonomy and theory. *Research Policy*, 13, 343–373.

Persson, U (1997). A conceptual and empirical examination of the management concept supply chain management. *Department of Business Administration and Social Science. Division of Industrial Logistics*. Luleå, Luleå University of Technology.

Pisano, GP (1997). *The Development Factory: Unlocking The Potential Of Process Innovation*. Boston, MA: Harvard Business School.

Rogers, J (2005). *Hot Commodities — How Anyone Can Invest Profitably in the World's Best Market*. Chichester: John Wiley & Sons Ltd.

Rosenberg, N (1982). *Inside the black box: Technology and Economics*. Cambridge: Cambridge University Press.

Sahay, BS (2003). Supply chain collaboration: the key to value creation. *Work Study*, 52, 76–83.

Samskog, PO, Björkman, J and Herbst, JA (1995). LKAB's Kiruna plant leads in developing plantwide process control. *Mining Engineering*, June.

Särkikoski, T (1999). *A Flash of Knowledge — How an Outokumpu Innovation Became a Culture*. Espoo, Finland: Outokumpu Oyj.

Schroeder, R, Van De Ven, A, Scudder, G and Polley, D (1986). Managing innovation and change processes: findings from the Minnesota innovation research program. *Agribusiness*, 2, 501–523.

Skinner, W (1992). The Shareholder's Delight: companies that achieve competitive advantage from process innovation. *International Journal of Technology Management*, 41–48.

Soosay, CA, Hyland, PW and Ferrer, M (2008). Supply chain collaboration: capabilities for continuous innovation. *Supply Chain Management: An International Journal*, 13(2), 160–169.

Spitz, PH (1988). *Petrochemicals: The Rise of an Industry*. New York: John Wiley & Sons.

Stobaugh, R (1988). *Innovation and Competition — The Global Management of Petrochemical Products*. Boston, MA: Harvard Business School Press.

Taylor, SG, Seward, SM and Bolander, SF (1981). Why the process industries are different. *Production and Inventory Management Journal*, 22(4), 9–24.

Tidd, J (2001). Innovation management in context: environment, organization and performance. *International Journal of Management Reviews*, 3, 169–183.

Tottie, M and Lager, T (1995). QFD — Linking the customer to the product development process as a part of the TQM concept. *R&D Management*, 25(3), 257–267.

Urban, GL, Hauser, JR and Dholakia, N (1987). *Essentials of New Product Management*. Englewood Cliffs, NJ: Prentice-Hall.

Utterback, JM (1994). *Mastering the Dynamics of Innovation: How Companies can Seize Opportunities in the Face of Technological Change.* Boston, MA: Harvard Business School Press.

Utterback, JM and Abernathy WJ (1975). A dynamic model of process and product innovation. *Omega*, 3, 639–655.

Webster (1989). *Webster's Encyclopedic Unabridged Dictionary of the English Language*. New York: Portland House.

Wittgenstein, L (1992). *Filosofiska Undersökningar*. Stockholm: Thales.

Woodward, J (1965). *Industrial Organizations: Theory and Practice*. London: Oxford University Press.

PART 2

STRATEGIC PROCESS INNOVATION

The message in Mintzberg's book *"The Rise and Fall of Strategic Planning"* is not that strategy and strategy-making are obsolete, but that they should be carried out in decentralized management teams and that the strategy should be an open and flexible document (Mintzberg, 1994). The philosophy advocated by Mintzberg is still a good platform for a dynamic strategy development approach. We are not talking here about superficial and badly compiled documents which after being written will gather dust on the bookshelves of company executives. On the contrary, it is advocated that every corporation needs an ongoing strategy work process where strategic issues are carefully worked out and intermittently reassessed. Because of the investment-heavy nature of the process industries and their production process technology, the need for such long-term strategic guidance of process innovation is unquestionable.

Companies in the process industries often have a history of making products using processes that sometimes originated hundreds of years ago. In today's global competition, however, the often tacit knowledge about how product functionalities are created and about associated production processes must gradually change in the future. There is thus a need for a deeper understanding, resting on more factual information and on a stronger scientific platform. The creation of such platforms needs short and long term objectives and plans for innovation, and one appropriate tool often used now for such an endeavour is Technology Roadmapping (TRM) with the development of associated Strategic Research Agendas (SRAs).

This part of the book will start with a discussion of the newness of process technology and process innovation. The concept of distributed innovation intensities are introduced, and the relationship between product and process innovation is then discussed. Further on, the use of the Quality Function Deployment (QFD) methodology for process industry applications (mpQFD) will be demonstrated as a tool to facilitate such a roadmapping exercise in the creation of a more fact-based platform. It is also an overall tool for progression of customer demands back to necessary development of process capabilities and future demands on raw material specifications. Finally, in the last chapter, TRM is presented as an important tool to link overall corporate business objectives and strategy more firmly to innovation strategy-making. The development of an SRA for innovation that is related to the roadmap will be discussed, and the development of a process innovation strategy addressed.

Chapter 4

Process Innovation in the Process Industries

"In today's increasingly competitive global markets, a firm cannot survive for long if it is shackled with inadequate capacity, run-down and poorly located facilities, and outdated and uncompetitive process technologies."

Hays *et al.* (1996, p. 97)

Is the development of new process technology still a worthwhile undertaking, when the object nowadays is often to reduce the company's own share of production in the overall supply chain? Has production now become a "necessary evil", as Skinner provocatively puts it in his praise of a process development strategy (1992)? How important is company process development in the process industries now and in the future, and what is an appropriate distribution of R&D resources between product development and process development? Although these questions are certainly not easy to answer, they have to be addressed by top management, and in particular by the director of R&D, in making strategic decisions for the future, or at least in the development of the company's annual R&D budget.

Customer orientation and improvement of company product development capabilities have been matters of concern in the process industries for several years, but what about process development? Do the sometimes short-term perspectives of shareholders, analysts and top company management allow too few resources for process development,

59

because they are afraid that the development will take too long and commit them to heavy capital outlay in the future? In Skinner's (1992) article comparing the process industries with "non-process industry" he asks: "What can be learned from the process industries, where process innovation seems to be a way of life?" The question is whether process development still is and should be a way of life, and as such an important area of R&D for companies in the process industries.

In the first section of this chapter the newness of process technology will be discussed in the perspective of a technology S-curve and a "Matrix for process innovation". Afterwards the relation between investments in product innovation and process technology will be treated in the light of one dominant model and research results from the ERP. Innovation in the supply chain in the process industries will be discussed in the last section of this chapter. The strong interconnection between product properties, process technology capabilities and raw material specifications has justified the treatment of all those areas together in this book about process innovation.

4.1 The newness of process technology and process innovation

In the discussion of process technology and process innovation in the process industries the question of the newness of process technology will be addressed. It has always been common in the process industries that operating plants have areas where new and old process technologies are used together. Often, even newly built plants, for different reasons, have this combination of old and new process technology. Consequently it is important to recognize that in different parts of the process chain, process technology requires a greater or lesser degree of R&D effort and a greater or lesser share of the total available resources for process innovation. Because of this it may be misleading to discuss how mature process technology is in different plants, and even more so in industrial sectors, since the total process chains in individual firms may be totally different and include process technology that is both old and new.

The development of the Pilkington float glass process contradicts the use of such a concept on an industry level as well (Utterback, 1994).

The radically new float-glass process technology was developed and used to produce the commodity product flat glass, and the development can thus be classified as a pure process innovation project. In the flat glass industry sub-sector the Pilkington company is a good illustration of how an industry is often composed of a number of firms using technology of different newness. For a further discussion of the development of the Pilkington process and the flat glass industry see Utterback and Anderson (Anderson and Tushman, 1990; Utterback, 1994). Using the maturity concept on an industry level is thus rather questionable, since firms within each industry may utilize different technologies with different degrees of newness.

The S-curve concept

One tool in the discussion of technology newness is the "S-curve" concept (Foster, 1986). S-curves of a process technology are drawn in Figure 4.1. A simple explanation of such an S-curve is that the

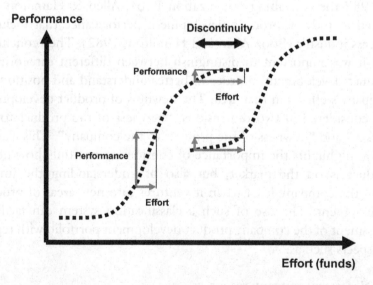

Figure 4.1   The S-curves developed after Foster. The effectiveness is set when a company determines which S-curve it will pursue. Efficiency is the slope of the present curve. Effectiveness is related to the strategy and the selection of a new technology trajectory (Foster, 1986).

introduction of a new technology often takes a lot of effort to get it going. Afterwards there comes a second phase when the performance of such a technology can be much improved with limited innovation resources. When the technology becomes more mature, performance improvements again tend to take more effort (Anderson and Tushman, 1990). At this stage, what Foster (1986) calls the "discontinuity", a switch to a different technology, may be advantageous — if not necessary to survival.

The S-curve can serve as a useful tool for firms in discussions of the maturity and competitive strength of unit process technology and a firm's main production technologies, especially during discussions of investments in new process technology. The use of this concept can be helpful in the development of a process technology roadmap and be a supplementary tool in the development of a corporate process innovation strategy, which will be further elaborated in Chapter 6.

What is new about a new process technology?

In 1982 the consulting organization Booz, Allen & Hamilton presented a study of product development performance that included process industry (Booz Allen and Hamilton, 1982). They concluded that it was important to distinguish between different categories of product development in order to better understand and position the company's efforts in that area. The newness of product development was considered in two dimensions, "newness of the product to the market" and "newness of the product to the company". This classification highlights the importance of considering not only how new a product is on the market, but also of understanding the impact upon the company itself when it ventures into new areas of product development. The use of such a classification system can facilitate assessment of the company product development portfolio with regard to aspects such as:

- Necessary company resources.
- A proper risk/reward balance for the product development portfolio.
- Personnel qualifications needed for different kind of product development.

The importance of a better classification of product development is gaining acceptance in industry, but the Booz, Allen & Hamilton product matrix has also been used in academic research and for the classification of different types of success criteria for product development (Griffin and Page, 1991).

A matrix illustrating the importance of different types of product and process development for different types of process industry has been presented by Cobbenhagen *et al.* (1990). The importance of process development to industries when producing commodity products, is recognized by Hutcheson *et al.* (1995). The pharmaceutical industry often still gives higher priority to product development compared to process development, but as the time frame for process development increases, the need to have product and process development going on in parallel to shorten lead times is emphasized by Pisano (1997).

The importance of process development in many industries belonging to the process industry group is recognized, but process development, like product development, varies widely in its nature and scope from incremental improvement to creation of highly innovative, radically different process technology (Freeman, 1990). There is consequently also a need for a good characterization of different types of process development and for an answer to the question "What is new about a new process?" Confusion is sometimes caused both in industrial R&D and in academic research when completely different categories and types of process development are compared to each other. The aforementioned study by Booz, Allen and Hamilton used the dimensions "newness to the market" and "newness to the company". The importance of newness to the company in the development of process technology has previously been pointed out by Linn (1984). Newness to the market, however, is not a relevant consideration for industries that are not developing new process technology to be marketed outside their own organization. Newness of the process technology to the world is probably a better dimension in this case.

Newness of process technology to the world

The newness of the process technology to the world is a dimension that characterizes how well known or proven the process technology is outside the company. Newness of process technology to the world is the traditional dimension for classifying different types of development and is consequently also an important dimension to select when classifying process development.

The degree of newness of a process technology can sometimes be related to whether the process can be patented, but since new processes are sometimes not patented but kept secret, the newness can also be estimated by how well it is described in professional publications. It is suggested that three degrees of newness of a process technology to the world should be distinguished (Lager, 2002b):

- Low: the process technology is well known and proven (can often be purchased).
- Medium: the process technology is a significant improvement on previously known technology (incremental process technology development).
- High: the process technology is completely new and highly innovative (breakthrough or radical technology development).

Newness of the process technology to the company

One thing that often distinguishes the process industries from other manufacturing industry is that production plants in the former are seldom easily modified, and changes in the process configuration are often costly and investment-intensive. Because of this, the newness of process technology to the company production system is another important dimension in the classification of process development.

There are several possible ways to define the degree of newness of a process technology to a company, but before a company starts a process development project, one of the most important considerations should be how easily the process technology under development can be implemented in the company production system. The newness of a process technology to a company can thus be measured by the extent to which the introduction of a new process will affect the production plant/production line/production unit in terms of investment in new production equipment or a completely new production plant. Newness of process technology to the company production system has been chosen as the most readily understandable and usable dimension for professionals in the process industries. It is suggested that three degrees of newness of a process technology to the company production system should be distinguished:

- Low: the new process technology can be implemented and used in existing process plants.
- Medium: the new process technology requires significant plant modifications or additional equipment.
- High: the new process technology requires a completely new process plant or production unit.

The process matrix

A process matrix has been constructed in which the two selected ordinal variables rank the newness of process technology in two dimensions, both having trichotomous scales (Lager, 2002b). This new matrix is presented in Figure 4.2, where the two dimensions define a ninefold

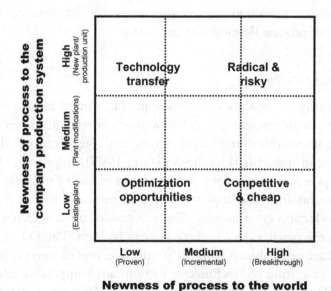

Figure 4.2   The process matrix for classification of the development of process technology in the process industries.

property-space. The matrix is then further reduced into four areas and labelled in this figure; this creates a simplified typology of process development projects that can facilitate use of the matrix for strategic project selection and portfolio balancing. The four areas in the matrix classify process development projects into four categories which are further discussed in the following.

Optimization opportunities

The use of proven or incrementally improved process technology in an existing plant environment might not necessarily create a competitive production system for the future, but may nevertheless be of significant importance in a short-term production perspective.

This type of process improvement is not the development of a new process, but the refinement and optimization of an existing process into a cost-efficient production system. It does not necessarily require any new equipment at all, but only minor changes in process conditions or in flowsheets and process configurations. Using mainly existing equipment and possibly new reagents, additives or raw material qualities reduces both the investment and the risk involved, making this type of process development attractive.

Technology transfer

Development of process technology in the process industry is to a large extent dependent on collaborative efforts and projects with equipment manufacturers, contractors and suppliers of chemical reagents and materials (Hutcheson *et al.*, 1995). Improving the production process in a company is thus often a matter of utilizing already proven technology and applying it in part or as a whole in the company production environment. The newness of process technology is low, but the newness to the company can be high. The risk of experiencing start-up problems with the production unit or new plant is low. The risk of getting old technology without any competitive advantage can on the other hand be high (the vendor of a turnkey plant is not likely to take undue risks). Allocating company R&D resources to this area should be considered with care, and process development should preferably take place in collaboration with external partners. For an in-depth discussion of technology transfer, see Chapter 12.

Radical and risky

In this area we have medium-to-high newness in both dimensions — a real process innovation and breakthrough technology that is new to the company production system. Process technology here is of a kind that can make old plants or production units obsolete (Cobbenhagen *et al.*, 1990). Such a new plant can create a completely new and highly competitive production system, but the first player is often taking a higher risk. Depending on the size of the process and necessary capital for the investment, this type of development project should be approached with caution if contemplated by a smaller company.

Competitive and cheap

This is an interesting position in the matrix where the degree of newness is medium to high in its newness to the world. Since this type of process development requires only moderate plant investment with minor new process equipment, the risk is low, but the profitability could be high. This is an attractive area for process development and an area where initiatives should possibly be encouraged. The development and introduction of improved process control systems is one example that position well in this area.

Evidence from the European research project

The following results are from the previously presented ERP, which is described in App. C (Lager, 2002). Six groups of industries were included in the study: mineral and mining, food and beverage, pulp and paper, chemical, basic metal and others. Figure 4.3 shows the distribution of average annual expenditures on development of

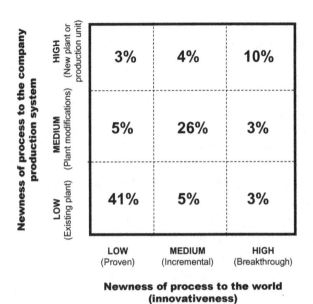

Figure 4.3 Company annual expenditures on development of process technology among areas of the process matrix. Mean values for all companies in the survey (Lager, 2002).

process technology in the total group of companies included in the survey. Out of the total annual expenditures on process development, 41 percent is spent in the area of proven technology/existing plant. This shows that the largest part of process development takes place in an area that could be called optimization of available production systems.

The second most important area, one that is often called process development by process engineers, is incremental process development/plant modifications, to which 26 percent of expenditures on process development are allocated. The third area in ranking order is breakthrough technology/new plant, to which 10 percent of the resources for process development are allocated. The first two of these areas add up to 67 percent and all three to 77 percent, leaving 23 percent spent fairly uniformly over all other areas of the matrix (3–5 percent in each area).

Because of lack of space, the distribution of annual expenditures on development of process technology for each of the six aforementioned categories of process industry is not included here, but the general impression from Fig. 4.3 holds good for all of them. Two categories — mining and mineral, and food and beverage — devote most of their process development expenditure (49 percent) to proven technology/ existing plant.

The minimum expenditures in this area (37 percent) are in the chemical and the steel and metal industries. If we add up the modified plant and new plant areas in the breakthrough technology column, we get a fairly high figure for both of these (16–17 percent), indicating a high degree of innovative process development. High standard deviations for all individual areas and for all categories of process industry indicate large individual variations within each category.

Summing up, the proven technology column in the matrix for all respondents shows that about 50 percent of all process development work is associated with implementation of already existing process technology. This stresses the importance of the R&D organization's ability to apply proven technology to existing production plants and production units. In the food and beverage industry, 63 percent of all process development resources are spent on development of process technology without any innovative input at all. There is a preference for the diagonal in the matrix, indicating that the more innovative the new technology is, the more changes in production plant are necessary for implementation. This association is plausible, but could also possibly be attributed to some extent to the terminology used in the matrix.

Since 80 percent of the respondents completed the matrix correctly, it is an indication that it was not considered too difficult or pointless a task, thus giving additional support to the usability of the matrix. It is interesting to note that those respondents who did not complete the matrix gave an even higher ranking to both importance and usability than the group that completed the matrix. This could possibly indicate that they did not have the figures at hand, but saw the need for them.

4.2 The relation between product and process innovation

Introducing the concept of distributed innovation intensities

Assuming that the importance of company process development is, or should be, reflected in process development's share of total annual expenditures on R&D, and remembering that the development of products and processes still often takes the lion's share of all R&D resources in the process industries, the balance between product and process development should be an important and strategic issue for R&D management (Lager, 2002a). The concept of "frequency of major innovations" (previously called "rate of innovation") used by Utterback & Abernathy (1975) as a measure of innovation in their models, is not well suited for use in industry.

The concept of R&D intensity, however, is well defined and often used when R&D expenditures are compared between different sectors or firms in industry (see Chapter 3). The need for comparative yardsticks for product and process development that can be related to time has given the incentive to formulate two new concepts. By analogy with the definition of R&D intensity, it is proposed that "product development intensity" be defined as a company's annual expenditure on product development divided by its total annual expenditure on R&D, expressed as a percentage (Lager, 2002a).

Similarly, "process development intensity" is defined as a company's annual expenditure on process development divided by its total annual expenditure on R&D, expressed as a percentage. The sum of product development intensity and process development intensity is not necessarily 100 percent, since there may also be other types of R&D activities going on in a company. Since company R&D resources are allocated not only to product and process innovation but also to a number of innovation and innovation-related activities, it is proposed that all innovation intensities of different kinds are denominated "distributed innovation intensities". A company's expenditures on the different kinds of innovation and innovation-related activities divided by its total annual expenditure on R&D, expressed as a percentage will thus be further used in this book and called distributed innovation intensities (see Figure 3.1).

The Utterback and Abernathy models revisited

In 1975, Utterback and Abernathy presented "a dynamic model of process and product innovation" where they related the rate of innovation in industry to its stage of development over time (1975). The model was tested with empirical evidence from a previous study (reference to Myers and Marquis, 1969). The original model was later discussed and developed into a slightly different form but with no further empirical evidence (Utterback, 1994).

As a complement to this further developed model, new "waves" of innovation are sometimes also included when later stages in industry development have been reached. The original model was further developed, in that a distinction was made between industries that manufacture "assembled products" and those that manufacture "non-assembled products" (see Figure 4.4 and the model for industries making "non-assembled products").

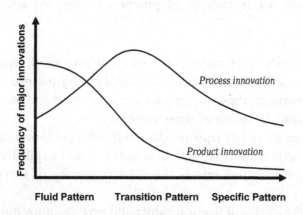

Figure 4.4 The developed Utterback & Abernathy model and patterns of innovation for non-assembled products. The frequency of major innovation in different phases of industrial development over time show different patterns for product and process innovation (after Utterback, 1994).

Three phases in industrial development were identified, classified and named in the original model (later renaming by the authors in parentheses):

- The uncoordinated (fluid) pattern. Product performance-maximizing strategy. Classified on the basis that most innovations are stimulated by market needs.
- The segmental (transitional) pattern. Sales-maximizing strategy. Classified on the basis that most innovations are stimulated by technological opportunities.
- The systemic (specific) pattern. Cost-minimizing strategy. Classified on the basis that most innovations are stimulated by production-related factors.

The evidence from the empirical testing of the original model seems to confirm the model on an industry level. A more fruitful approach than discussing whether or on which level of analysis those different models are valid, and whether they are descriptive or prescriptive, is to highlight the important aspects of product and process development that Utterback & Abernathy put forward in their models, and to use them for further development of improved concepts and models:

- The frequency of major innovations is presented separately for product and process innovation, highlighting the need to distinguish between these two types of development and also the need for a clear definition of those concepts.
- The frequency of major product and process innovations is indirectly related to time, and although it is not explicitly stated in the model, it also relates the total frequency of innovation to time.
- The frequencies of major product and process innovation are presented and discussed together. This focuses attention not only on the frequency of each kind of innovation, but also on how they are interrelated.

- It is proposed that different models should be applied to the process industries and other manufacturing industry, recognizing the importance of distinguishing between the two categories of industry and the need for clear definitions of those categories.

Empirical evidence from the European research project

How important process development is, and consequently how much of company R&D resources is allocated to process development in the process industries now and in the future, was one research question to the respondents in the European Research Project (ERP), further described in App. C. Process development intensities were given directly by the respondents at the beginning of the questionnaire (they were asked to state the share of total company R&D spent on process development). The results are presented in Fig. 4.5, using box plots where the bold horizontal line in each box shows the median, the distance between the ends of the box the interquartile range (50 percent of all values), the whiskers the highest and lowest figures. The arithmetical averages are also shown in parentheses in the following text.

The median and arithmetical averages of process development intensity for the total group of companies in this study are both 40 percent, with an interquartile range of 30 percent. It can thus first of all be noted that there is a tremendous spread in the figures for the process development intensity for the total group of companies in this study: the process development share of total R&D expenditures ranges from 0–100 percent. An ANOVA test shows significant differences in process development intensity between the different categories of industry ($p < 0.001$). A post hoc test with Bonferroni correction shows significant differences between two groups. The first group comprises the three categories with a median figure around 50 percent. The highest median value among the industries is 55 percent (mean 55.5 percent) for mining and minerals, followed closely by 50.0 percent (38.9 percent) for pulp and paper, and

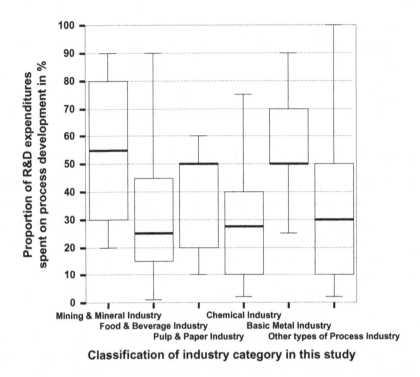

Figure 4.5   Process development intensity in different categories of the process industries, presented as box plots where the bold horizontal line in each box shows the median for the category. The box represents the interquartile range. The whiskers show the largest and smallest observed values.

50.0 percent (57.1 percent) for basic metal. The other group includes the three other categories with a median close to 30 percent. This includes other types of process industry 30.0 percent (36.1 percent), chemical 27.5 percent (27.8 percent), and food and beverage 25.0 percent (33.3 percent).

No category of process industry has closely grouped values, and the spread is very high within each category. As shown in Fig. 4.5, the mining and mineral industry has an interquartile range of 54 percent (the range where 50 percent of all values can be found). For the food and beverage industry the interquartile range is smaller but the range for the lowest and highest value is nearly 100 percent. The strange-looking boxes with medians at the top or bottom represent groups of companies with a large number of values identical to the median and this quartile. If we look at the figures from an opposite angle, using the previously discussed assumption that product and process development intensities add up to approximately 100 percent, product development intensity is highest in the food and beverage industry at around 75 percent, compared to mining and minerals where product development is the lowest at around 45 percent.

In this study, 40 percent of total company R&D expenditure, as an arithmetical average for all companies, was spent on process development. The answer to the question of whether process development is important in the process industries is on the whole affirmative. If we assume that all other company R&D resources are spent on product development (which however is not certain in the light of the previous discussions and definitions), the complementary figure for product development intensity is 60 percent.

This figure indicates that the importance of product development in this group of companies is currently higher than, or at least as high as, that of process development. The figure is at first sight somewhat surprising, but seems plausible since companies in the process industries may have focused more on product development over the past 10 years (a comment based more on anecdotal information from the author's industrial experience than on any specific research results). The results show a remarkably high dispersion within each category of industry. This may be an indication that there are more factors influencing the relation between product and process development intensities than only the type of industry, which was selected as a discriminating variable for the data analysis. The primary reason for selecting type of industry as a variable is that it refers to industrial reality and is related to the competitive environment in which each industry operates.

One explanation of the spread in the results could be that individual companies are currently in different phases of product and process development (which was also pointed out by respondents in the study) and that the relationship between product and process development does not necessarily follow a generic homogenous pattern for a sector of industry as a whole, but a company-specific pattern which may or may not be cyclic in nature. Another explanation could be that the different companies have different customers whose needs prompt them to adopt different overall corporate strategies. This might require different emphasis on R&D and consequently on R&D intensity. Different R&D strategies and variations in R&D intensity could then emerge as different patterns for both product and process development intensity.

As a follow-up to the question about process development's share of total R&D expenditures (process development intensity), respondents were asked about the future prospects for company process development: "Do you estimate that the process development share of total development work will decrease, not change or increase in your company?" The results are presented in Table 4.1.

The results show an overall trend towards an increasing share of process development in the future. This is very clear, around 80 percent, for the industry group including mining and minerals, and food and beverage, but here we also find the two highest figures for a decrease. In other words, change is expected in both directions. The other group comprises the remaining categories, and in this group the chemical industry has the lowest figure for an expected increase in process development at 53.6 percent. The results clearly show that an increase in process development intensity is expected in the future by most of the companies in this study.

One explanation of this rather interesting trend could be that the emphasis on product development during the past 10 years has resulted in too little attention being paid to process development, even to the point of neglect, and that the time has now come to focus more on this area. Another explanation could be that if spending on product development and new products has increased, that very fact will tend to encourage process development. The explanation of the fact that two groups of process industries show slightly different patterns for the expected increase of process development intensity might be that companies in the mining and mineral industry often tend to be commodity producers with a need for more process development (see further Figure 2.3), while the food and beverage

Table 4.1 Expected Change in Process Development Intensity for Different Categories of the Process Industries. Percentage within classification of industry category in this study.

	Classification of industry category in this study						
	Mining & mineral %	Food & beverage %	Pulp & paper %	Chemical %	Basic metal %	Other types of process %	Total %
Decrease	9.1	11.1			4.3	4.3	4.5
No change	9.1	11.1	33.3	46.4	39.1	34.8	32.1
Increase	81.8	77.8	66.7	53.6	56.5	60.9	63.4
Total	100.0	100.0	100.0	100.0	100.0	100.0	100.0

industry currently has the lowest process development intensity, which possibly needs to be increased for community production.

Comparing the results with the previous model

These results do not fit the Utterback and Abernathy model. A further development of the Utterback & Abernathy model gives a relationship between the frequency of major product and process innovation for non-assembled products in different stages of industry development (Figure 4.4).

They discuss the model on industry, company and product category levels. In the initial Fluid Pattern, product innovation is twice as frequent as process innovation; in the following Transition Pattern, process innovation is at least twice as frequent as product innovation; and in the final Specific Pattern, process innovation is about five times as frequent as product innovation (Utterback, 1994).

If one assumes that most of the industries and companies in this study can be classed as so-called mature, they should according to the Utterback & Abernathy classification be in their last phase of industrial development, follow the Specific Pattern and have much more process development than product development. While one cannot claim that the concept of "frequency of major innovations" equals the concepts of "product and process development intensity", it is as close as one can get to a comparison of the theoretical model with industrial reality. If on the other hand the companies are not considered mature, or if different companies are in development phases corresponding to the Utterback and Abernathy "new waves", the result might fit the models, but then on the other hand all possible results fit the models, and the models are not testable and of minor interest.

One could also argue that the expected increase in process development's share of total R&D resources could fit the Utterback & Abernathy model, with process development becoming progressively more important than product development in mature industries making non-assembled products. But on the other hand, product development intensity is higher than process development intensity in the total group of companies in this study. If we look at the individual cases in the study, there do not seem to be patterns that can be explained by different waves of development, and the conclusion is that the results from this study do not support the models. Other studies, e.g. Pisano (1997) and Bower & Keogh (1996) have also observed this lack of fit with the model and empirical data from industry.

4.3 Process innovation in the supply chain

Innovation in the process technology downgrade/upgrade cycle

Technology planning can serve many purposes in the process industries. It will first of all give a good hierarchic structure of technologies and provide a framework for an assessment of the company's process technology capabilities. Further on, it can be used as a framework for possible outsourcing of competence and a platform for decisions about external process innovation collaborations and partner selection.

In the technology planning process and in the further creation of a process technology roadmap, it is often advantageous to classify technologies in the following categories (Dussauge *et al.*, 1987, Granger, Hamel and Heene, 1994):

- Basic process technology: Essential knowledge for the business, widely exploited by competitors, with little competitive impact, and already generally available on the market at a cost.
- Core process technology: Key proprietary process technology with strong competitive power, not well known or efficiently used by competitors.
- Emerging process technology: Not available on a supplier's shelf, but likely to have a future competitive impact. Emerging technologies can be of different kinds. Nanotechnology and biotechnology can now be considered as emerging technologies in some industries. Technologies not used in a firm's own industry sector but in other non-competing industry sectors may also be classified as an emerging technology.

In the internal benchmarking of technologies, the following nine degree scales can be deployed:

Importance of the technology to the company

- 9 = Of extreme importance
- 7 = Very important
- 5 = Important
- 3 = Less important
- 1 = Unimportant

Company's technological position

- 9 = Global leader
- 7 = Strong position
- 5 = Favourable position
- 3 = Difficult to keep up competence
- 1 = Weak or non-existent

In technology planning, core process technologies are identified, and strategies for their development, use and protection can be formulated. Basic technologies are more a matter of finding the best collaborative partners for selected technologies. Core corporate process technologies will, in the long run, degrade into basic process technologies, and how to go ahead in the degradation/upgrade cycle is a strategic corporate decision (Fig. 4.6).

The boxes on top of and under the production process illustrate the dependence on external providers of production technology/equipment and knowledge that are not available in-house in the company. This will be discussed in Part 4. In this figure, the previously presented downgrade/upgrade cycle for products has been supplemented and extended backwards in the supply chain. Two new simplified models, the downgrade/upgrade process technology cycle and the downgrade/upgrade cycle for raw materials, are introduced.

Lacking sufficient resources for process innovation, corporate core process technology will inevitably and in due time be downgraded to basic technology. The same goes for the raw material supply, when lack of resources for raw material development will impoverish the company's raw material supply base.

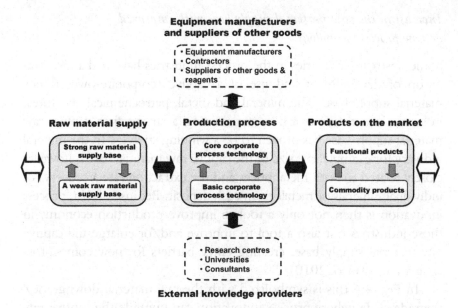

Figure 4.6 A new conceptual model of the downgrade/upgrade cycles for raw materials, process technology and products in the process industry supply chain. The quality of raw material supply will not only influence the complexity of necessary process technology but often also the final quality of delivered products. Process innovation will interact in both directions in this total supply chain, which may also extend further in both directions.

Innovation in the raw material downgrade/upgrade cycle

Securing the supply of raw materials for the production process has always been an important aspect for firms operating in many sectors of the process industries. The latest raw material business super-cycle has proved this point well. Raw material properties will not only influence production costs (higher or lower) and the complexity of appropriate production process technology (more or less complex), but also often determine the quality and the performance of final products (higher or lower performance); see Fig. 4.6. The specifications for the raw materials going into the production process are then only one side of the coin: the evenness of each specification is the other side.

Increasing the raw material supply base with improved or new process technology

Some upstream industries in the process industries have had a long tradition of development and use of a captive (corporate-owned) raw material supply base. The mineral and metal, petrochemical and forest industries are examples where maintaining a strong firm captive raw material supply base has proven to be a winning strategy. In the mineral industry for example, the size of the captive raw material base is however closely related to both metal prices and the extraction rate (recovery) of individual minerals or metals from each deposit. Because of that, process innovation is then not only a tool to improve production economy in those industries but also a tool to improve and/or enlarge the captive raw material supply base, creating entry barriers for new competitors (Lager and Blanco, 2010).

In Fig. 4.6 this is symbolized with the raw material downgrade/upgrade cycle, where process innovation can upgrade the captive raw material supply base. In some of the "upstream" process industries (or the upstream part of a firm's internal supply chain), improved exploration of existing deposits or search for "green field" deposits is however the normal way to upgrade the captive raw material supply base. Such exploration will, on the other hand, often benefit from close internal collaboration and interaction with process innovation in order to find innovative process and extractive technology solutions for improved raw material utilization.

The supply of purchased raw materials and other commodities

Companies located further downstream in the material supply chain need to secure suppliers of price-competitive raw materials with desired material specifications and processability. An external collaborative approach in the development of such raw materials is an avenue that has not been explored to its full potential so far. It is thus really surprising that securing a price-competitive and high-quality raw material supply base in a long-term perspective, and the collaborative development of such raw materials, still have not been given a higher priority by companies in the process industries. In other manufacturing areas, for example the automotive industry, an intimate collaborative innovation with component suppliers in the supply chain has long proven to be a successful business strategy (Miyashita and Russel, 1995). Suppliers taking over the development of improved components and systems is a collaboration model that was born in Japan but now has gained a world-wide acceptance. The collaborative development of raw materials with suppliers can still be regarded as being in its infancy in the process industries.

The use of recycled materials substituting for other raw materials is also likely to be of ever-increasing importance in the process industries in the future. Virgin raw materials will then gradually be replaced by recycled materials; scrap-based metallurgy, for example, is becoming more important in the steel Industry, and recycled paper products in the forest industry. This trend has also been influenced by external regulations. Further process innovation and product innovation will then determine what in the future is to be regarded as a recyclable resource for creating new sources of raw materials. The optimization of the total raw material supply chain, including an improved functionality of other commodities and reagents for the production process, is an area with large, but probably often neglected, economic potential.

It is advocated that the innovation area "Development of raw material supply" which was introduced in Fig. 3.1 in Chapter 3, deserves to be better recognized as an important area for innovation in the process industries. The strong interaction with process innovation and even product innovation is the reason why it also should be included in the development of a corporate roadmap and a strategic research agenda for innovation.

4.4 Summing-up and some issues to reflect upon

The concept of maturity, specifically the maturity of industries, is initially discussed. The conclusion is that it is not proper to use this concept in relation to industry or sectors but more on a company, and certainly on a technology, level. The performance of process technology can however be related to development, using the S-curve concept, where the results of process development are given as a function of the required development effort. It is argued that a simple classification of technologies using the S-curve concept is one important piece of information in the analysis of the need for process innovation.

The newness of process development is later discussed using the process matrix. This matrix underlines the importance of viewing corporate innovation from the perspectives of both newness to the world and newness to the company. Different areas in the matrices are named and explained. It is advocated that each company should position either its various projects or its total allocation of resources for process innovation in the matrix, and that using different symbols gives a good overview of the strategic deployment of the innovation resources. If a company is looking for breakthrough process technology, it will not get that if its resources are allocated to incremental process improvements. The results from the ERP show that a fairly small amount of available resources were allocated to process innovation of high newness, a state of affairs that should be carefully evaluated in the formulation of firm process innovation strategies for the future.

The new concept of "distributed innovation intensities" is introduced and further discussed and is used in the definitions of "process innovation intensity" and "product innovation intensity". The allocation and distribution of innovation resources to product and process development is discussed, starting with the classical Utterback & Abernathy model. The analysis of the model provided important information for investigation of the state of affairs in the European process industry. The results from this study show that there is a wide dispersion of process innovation intensities between different industrial sectors and also within each sector.

In the last part of this chapter it is pointed out that there has been too much focus on the products on the market and product development, and too little focus on how to develop competitive production technologies to produce those products in a competitive efficient production system. Not only that, the issues of how to secure and develop cost-competitive raw material supplies and the further development of the processability of such materials have been grossly neglected in research into technology management.

Forward integration has long been the mantra in new strategy development but what about backward integration? In resource development, collaborative development with raw material suppliers can be looked upon as the reverse image of application development, an area that will be thoroughly discussed later in Part 4.

- *Select the three most important process technologies in your firm and try to position them on a technology S-curve.*
- *Position the company's process innovation projects in the process matrix. Use different-sized circles to illustrate different project volumes and different colours for different project duration (short-term, mid-term and long-term).*
- *How important is production process technology to your company? How competitive is this process technology, and how much of this technology could be classified as "core process technology"?*
- *How well is process innovation defined in relation to product innovation in your company, and is process innovation getting its fair share of recognition?*
- *Are the resources allocated to process innovation sufficient and long-term enough to provide future competitive production process technology?*
- *Is the development of the company's raw material supply recognized as an important area for innovation? How much of total R&D resources are allocated to this area?*
- *How much of the company's raw material supply is "captive" and if so, what is the estimated lifetime of the captive reserves?*
- *In a dynamic perspective, how much would those captive resources increase with 20 percent improved raw material recovery in the process, or 20 percent lower production costs or 20 percent higher market price for finished products?*

Chapter 5

Multiple Progression Quality Function Deployment — from the Voice of the Customer to Process Capabilities and Raw Material Specifications

"QFD appears complicated at first glance, and technical personnel might tend to respectfully ignore it, but the data can be considered as an accumulation of the past that can be added to or improved with each new development cycle and therefore becomes an important asset to the company."

Nakahita Sato, former Director of Toyota Auto Body
(American Supplier Institute, 1989)

In the creation of a technology roadmap (TRM), which will be further treated in-depth in the following Chapter 6, one of the most difficult parts is to identify future fact-based customer needs and product targets as well as relating those targets to future necessary process capabilities and raw material specifications. How does one identify those future targets without relying on guesswork or old company market and production conceptions, and how to select and pinpoint important R&D areas in Strategic Research Agendas (SRAs) in order to meet those targets? It is argued in this chapter that the development of sound customer-driven roadmaps and the further development of robust associated SRAs can be facilitated in the process industries by using Quality

Function Deployment (QFD) and the Multiple Progression QFD system in particular (Lager, 2008).

QFD is a methodology and a work process where stakeholder (customer/consumer) demands on a product, process or a system are initially collected, refined and structured. They are afterwards evaluated (by ranking, rating or an analytic hierarchy process) by the stakeholders and benchmarked against competitors: the Voice of the Customer (VOC). The demands are afterwards related in a matrix to measurable properties on the product. The stakeholders' importance ratings are then translated into this new dimension where a technical benchmarking is performed and new target values are set; the House of Quality. The measurable product properties and their importance ratings and target values can then be further progressed as demands on the production systems and further backwards into raw material specifications: Multiple Progression QFD.

After a short introduction to the QFD methodology, the newly developed QFD system for process industry applications is presented. The use of the system is then illustrated by a tentative application of the system in the cement industry.

5.1 An introduction to quality function deployment

Quality Function Deployment (QFD) originated in Japan, where quality work had been carried out from the beginning of the 1950s. It had its breakthrough in Japanese manufacturing industry, and is often attributed to Toyota Auto Body, whose use of QFD successfully contributed to solving their problems with rusting cars in the mid-1970s. The first and original system, developed by Professor Akao (1990), is still the dominant system used in Japan and was introduced into the USA in a slightly modified form by the GOAL/QPC company (King, 1987). This system includes a large number of matrices and is often referred to as the "Matrix of Matrices". A more simplified version developed by Dr Fukuhara received wide publicity when it was presented in the Harvard Business Review under the name "House of Quality" (HOQ) (Hauser and Clausing, 1988), while the stair of four consecutive matrices often is referred to as the "Four Phases of Matrices". QFD was introduced at the beginning of the 1980s in the USA and at the end of that decade in Europe.

QFD is unfortunately often presented in a rather superficial manner using simplified descriptions of the HOQ and finishing with a presentation of how customer demands are progressed to production planning using four consecutive matrices. The methodological reality is however more complex, since there are first of all two different systems in use today (Martins and Aspinwall, 2001). The four-stage progression system is not however applicable in the process industries because assembled products are not produced (Lager, 2005). The translation of customer demands (Whats) into product properties (Hows) takes place in the HOQ in both the Matrix of Matrices and Four Phases of Matrices systems. The importance ratings of the hierarchically arranged customer demands, including comparison with competing products in the customer dimension (benchmarking), are usually collectively called the "Voice of the Customer" (VOC).

The Relationship Matrix is used to translate the VOC into an engineering dimension, developing measurable product properties. This includes identifying the direction in which individual properties should be developed with a view to pleasing the customer, as well as calculating the importance of individual product properties. In the engineering dimension there is now an opportunity to run a technical benchmarking of product properties. After completion of those "rooms" and individually designed rooms that are applicable to an individual project, the target values for a new or improved product can be set after a thorough matrix analysis.

Quality function deployment in the process industries

Successful future product development in the process industries will call for efficient systems for managing customer and consumer demands and translating this information into new or improved product and process concepts.

Such systems must match companies' short and long term development needs and must be adapted to their different types of customer and consumer structures. In combination with other work practices such as value engineering, QFD also has the potential to become an important tool for achieving quality and cost in product development (Hoque *et al.*, 2000).

The size and complexity of today's industrial organizations has sometimes made the distance between production, development, marketing and the customer so great that company functions have difficulty in clearly "hearing the Voice of the Customer" and disseminating it further within the organization. The VOC, including the delineation of individual customer demands and their total structure, rating their relative importance, and finally estimating how well they are satisfied in relation to competing products, is likely to continue to be a foundation for sound product development (Akao, 1990; Akao, 2003; Mizuno and Akao, 1994).

However, future product development in the process industries does not only require efficient systems for managing customer demands and measuring customer satisfaction: there is also a need to develop efficient systems to translate those demands into innovative product concepts (Pugh, 1981), defined by measurable product specifications. Linking those product specifications to selected process technologies and production control systems, and the further progression of end-user demands backwards to final raw material specifications, will be the hallmark of future product development of excellence in the process industries (Lager and Kjell, 2007).

The need for a QFD system for food industry applications has been pointed out by Benner *et al.* (2003). However, a simplistic use of QFD is not likely to reap the fruits of the methodology; use of the deeper levels of the systems and downstream matrices is thus necessary (Jiang *et al.*, 2007; Scheurell, 1994; Zheng and Chin, 2005). The need for a QFD system for process industry applications was pointed out by Lager (2005b), after a thorough review of QFD use in the process industries (see for example de Oliveira *et al.*, 1996; Hanson, 1993; Kubota, 1990; Ootaki *et al.*, 1996; Scheurell, 1992; Stitt and York, 1993).

5.2 Multiple progression QFD

A new QFD system was consequently developed based on those needs and was designated Multiple Progression QFD (mpQFD) (Lager, 2005b). The different matrices that are included in this system (Figure 5.1), and expected outcomes from implementing and using the system, are presented briefly in the following.

Customer understanding matrix 1A

The first component in the new system, Product Matrix 1A, has the complementary designation "Customer Understanding matrix" because its primary object is to translate customer demands (Whats) on the product further into a company engineering dimension and measurable product properties (Hows). All the rooms that have been conventionally used in a HOQ in other systems can also be used in this matrix (Figure 5.1). Importance ratings from the individual and hierarchically arranged customer demands are recalculated into importance ratings for individual product properties, using the relationship matrix. New or modified target values (new product specifications) are set, to be met in further short-term and long-term product development. Expected outcomes from the matrix analysis are the following:

Accessing end-user information

A structured approach to establishing contact and collaboration with the end-user in order to understand present and future needs with regards to existing or improved products.

Aggregating reliable customer (consumer) information

Systematically listening to the VOC for spoken or unspoken demands and expectations — not relying on traditional company perceptions or old market information.

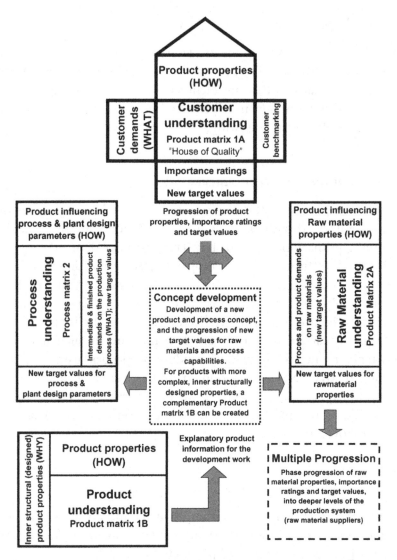

Figure 5.1   The simplified structure of the mpQFD system: the total system and its individual matrices (components). Starting with the HOQ, a number of matrices can be separately selected in the further progression, or alternatively all four of them. The large boxes with solid outlines are the four matrices (components) in the system, while the dotted box indicates concept development. The Multiple Progression of raw material properties can now be carried out in an infinite demand chain backwards in the production process (Lager, 2005b).

Progressing customer demands into the production process

Progressing consumer/customer demands backwards into the production process, further back along the internal and external demand chain (Rainbird, 2004).

A framework for new product concept development

Creating new product and process concepts is an important step in the development of new or improved products. Using present and new target settings from the different matrices give a firm foundation for this kind of activity (Pugh, 1981).

A platform for application development

In the process industries, the importance of application development is often second to none. The Customer Understanding Matrix provides a good starting-point for application development, and the development of application-oriented test methodologies.

Process understanding matrix 2

When individual product properties and their associated importance ratings and target values are progressed into the production process, they can be related to process parameters and raw material properties. Since process parameters, unlike raw material properties, cannot be progressed into deeper levels of the production process, and because processability can pose demands on the raw materials but not on process parameters, they have been split into two separate matrices in the new system. Finished product demands (Whats) for Process Matrix 2 are thus related to product-influencing process and design parameters. In this system the recommendation is to include the related process control matrix rooms directly under the relationship matrix related to the Hows of Process Matrix 2 to simplify the number of necessary matrices. Expected outcomes from the matrix analysis are the following:

Production process understanding

In the Process Understanding Matrix, measurable product properties are related to process control and design parameters. The completion of the relationship room will create a deeper understanding of the process and provide a fact-based platform for necessary process development and new process technology.

Linking product development to process capabilities

An efficient product development process should in the early phases often integrate the development of new or improved process technology. The Process Understanding Matrix is an excellent tool for analysing the necessary demands which new or improved product properties will make on the production process. The importance of creating product concepts in this matrix is matched by the importance of creating new "process concepts".

Functional cost analysis of homogeneous products

One prime objective of process development is to lower production costs. The cost of manufacturing the product will influence price, profit margin and product profitability. The Process and Raw Material Matrices can link production cost structure to product functionality, and be one tool for target costing.

Product understanding matrix 1B

An enhanced product understanding can be obtained through the development of the complementary Product Matrix 1B. In Figure 5.1, selected product properties from the Customer Understanding matrix 1A (Hows) and newly developed measurables of customer demands previously not measured are progressed into Product Matrix 1B, relating them to the products' inner (structural) properties or other explanatory measurables, called Whys.

When the Whats in the previously presented Customer Understanding Matrix are reviewed, and if it is necessary to develop new or more sophisticated measurement methods and techniques that capture those customer needs, those complementary Hows should be included in this matrix. If there are product properties that are not measured on a continuous basis in the production process (or if they are of a research character), this is the matrix where they should be introduced. Expected outcomes from the matrix analysis are the following:

Development-related applied research

The new Product Understanding Matrix is specifically designed to provide a strategic research agenda for necessary applied research. In the future, the knowledge base that is necessary for creative product and process innovation will need support from applied research, since the easier paths to new or improved products and processes are already trodden. Applied research then creates a good platform on which to base development activities. Completing all other matrices, and specifically the relationship rooms within the matrices, often also gives ideas and a guide to applied research in the creation of new breakthrough product functionalities.

Raw material understanding matrix 2A

In Figure 5.1, process and product demands (Whats) for Product Matrix 2A are related to raw material properties. The finished product demands are the ones that have been progressed from the previously presented product properties in the Customer Understanding Matrix. The process demands are the inherent demands the production process makes on the properties of raw-materials. These demands are not derived from the HOQ, but are collected from within the company's own production organization, and are designated processability demands. They can include demands on material handling properties as well as those related to ease of processing. The characteristics of measurable raw-material properties must be developed by judging the influence on finished product or processability. Those properties include the specifications that are given to the purchasing organization to be negotiated with raw-material suppliers in their management of the demand chain. Expected outcomes from the matrix analysis are the following:

Captive (strategic) raw material development

For companies in the process industries, an internal supply of captive raw materials is often an important competitive factor. The development of raw material often fails to receive the attention it deserves. The Raw-material Understanding Matrix gives important guidelines for this type of development.

Development with raw-material suppliers

Development in collaboration with external raw material suppliers in the process industries is in its infancy, compared to other manufacturing industries. Creating the Raw Material Matrices (one for each material) in collaboration with the suppliers is a rewarding exercise for both parties and gives a framework for sound raw material R&D agendas.

The usability of the system and brief outcomes from a case at Arla Foods

In a meta-analysis of nine scientific studies of the usability of QFD, the results definitely dismissed the old myth that QFD will give shorter time-to-market in product development. The good news was that one can expect to develop a better product that will give customers better overall satisfaction! In addition, it shows that the methodology gives several intangible benefits especially in the area of better communication and retrievable information (Lager, 2005a).

The results thus support the use of QFD not only as a tool for product development but also as one for accumulating information and for capturing the needs for both short-term and long-term R&D. These findings are supported by other studies that also emphasize the need for more customized applications and more "lean" QFD implementation (Ginn and Zairi, 2005; Herrman *et al.*, 2006). In an overall assessment of the first industrial project using mpQFD, the project was top ranked in the yearly corporate audit (Lager and Kjell, 2007). When the pros and cons of mpQFD were analysed in the internal evaluation of this project, the conclusions were:

- The system fitted this project and project objective very well, and the four matrices linked all the information together and created a backbone for the project. Information gathered in the project was collected, structured and assessed in a systematic manner.
- The matrix system succeeded very well in assuring that all project information was linked and related to actual consumer preferences — and not what Arla Foods believed were consumer preferences.
- Implementing the total Multiple Progression QFD system and developing good matrices is costly in terms of time and money, but what you gain is a very good general view over the complex relationship between consumer demands, product properties and how they can be manufactured and controlled.

The new system has thus proven to be a useful tool and a framework for the development of a wide spectrum of innovation activities generally performed in an R&D organization in the process industries, such as product, application and process development. For a full presentation of the case study, see Lager & Kjell (2007).

5.3   Using mpQFD as a tool for roadmapping and further strategy creation: A tentative explanatory application for the mineral industry

The cement industry is a part of the mineral industry and big international players today are Heidelberg Cement (Germany), Lafarge (France) and Cemex (Mexico), of which the first two are very large international companies with global businesses.

Managing the product supply chain for production

Figure 5.2 gives a simplified presentation of cement manufacturing, where large companies have integrated forward, and produce not only cement but also concrete for the construction industry.

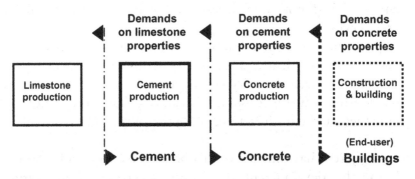

Figure 5.2   A simplified presentation of the Product Supply Chain and Customer Demand Chain in the cement industry. The shaded squares represent a forward-integrated cement corporation including in-house concrete manufacturing.

Cement and concrete production

Raw materials needed to manufacture cement are generally extracted from rock, mainly limestone. In the quarry the rock is blasted and crushed, then transported to the production plant. After the rock is ground to a fine powder it is sent to a kiln where the material is heated to around 1500°C. The finished product, called clinker, is finely ground after the addition of selected additives which enhance cement properties such as permeability, resistance to sulphates, improved workability etc. Concrete is a mix of water plus cement plus aggregates and additives. The formula may look simple, but modern concrete manufacturing is based on advanced scientific knowledge.

Aggregates are important ingredients in concrete, and influence its properties to a large extent. As a result the production of aggregate for concrete is an advanced technical production process including the unit operations crushing, grinding and screening. Concrete is now customized; with different additives such as water reducers, water repellents, air-entrainers and super-plasticizers, it can be given optimal properties for each application in the construction industry.

Managing the customer demand chain for innovation

The supply-chain perspective on production has been a very important advance in manufacturing, and as such has often resulted in dramatic cost savings and simplified production structures. Nevertheless it must never be forgotten that customers' perceptions in the future must be the drivers for sound corporate product and process innovation and are an equally important area for a competitive corporation (Rainbird, 2004; Walters, 2006). Managing the demand chain is an often neglected area, and is a joint responsibility of marketing and R&D.

In the process industries those demand chains are often long and complex, but must not be broken until they end in specifications for raw material properties (Figure 5.2). The mpQFD is a suitable tool for progressing end-user demands on key products to raw material specifications, further illustrated in Figure 5.3.

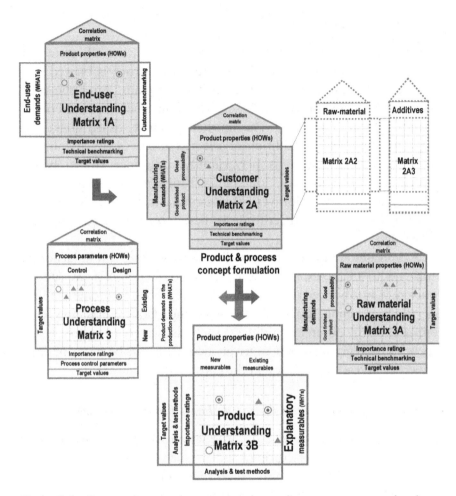

Figure 5.3 From end-user and customer understanding to process control and raw material specifications. An explanatory case of using mpQFD as a tool for managing the customer demand chain and as a platform for the development of a technology roadmap and a strategic research agenda. The arrows indicate progression of matrix data between matrices. The black arrows represent true multiple progression, from end-user to raw materials (Lager, 2008).

Starting with the contractor's demands on the concrete used in the construction process, in the End-user Understanding Matrix 1A, those demands are translated into measurable product properties of the finished concrete product in the relationship matrix (Lager, 2008).

Calculated importance ratings and new target values are further progressed as demands on the cement product in the Customer Understanding Matrix 2A.

Demands on cement properties are progressed from Matrix 1A, plus the demands on the workability of the concrete, here called processability. Those demands can also be further progressed into the matrices for aggregates and concrete additives (2A2 and 2A3). Understanding customer demands is a waste of time if they are not progressed into the manufacturing process. This is illustrated in Figure 5.3, where the measurable product properties of cement have been selected for further progression into the manufacturing process for cement; the Process Understanding Matrix 3. The product demands are also progressed further into the Raw Material Matrix 3A, where the relationship matrix translates those demands into limestone properties.

5.4 Summing-up and some issues to reflect upon

The Quality Function Deployment methodology is introduced and it is advocated that this is an excellent tool for capturing the VOC and translating it into measurable product specifications. The newly developed QFD system for process industry applications is presented and the benefits with the individual matrices are discussed in detail.

The usability of the mpQFD system in the process industries has been proven when utilising the system successfully in the food industries. The importance of understanding the supply chain and the customer demand chain for progressing customer demands back to raw material specifications was highlighted.

As a supplement to the traditional supply chain perspective, a demand chain perspective is introduced connecting targets and benchmarks for products, process technology and raw materials. The overall usability for roadmapping is illustrated with a tentative explanatory application in the mineral industry.

Further on, it is pointed out that in order to give guidance for the development of long-term process technology, there is a need to develop roadmaps not only for process technology, but for products and raw materials as well.

- *How does your company currently collect customer information on products, and have you ever considered using the QFD methodology on key products in order to listen better to the Voice of the Customer?*
- *How are future targets for product innovation presented to the R&D organization today in a competitive customer and technology perspective?*
- *How are necessary capabilities for future production process technologies for future products communicated backwards to the R&D organization for process innovation?*
- *How are different demands for the development of future efficient production process technology, originating from the production organization, articulated to the R&D organization? Do you listen to the internal customer, production, for process innovation?*
- *How does your company address its specific needs for the development of its captive raw material supply, or for a more efficient utilization of purchased raw materials, in terms of more adapted and efficient process technology?*
- *Is the long-term importance of efficient process technology in the total chain of raw materials, process technology and product performance recognized in your company?*

Chapter 6

Technology Roadmaps and a Strategic Research Agenda for Process Innovation

"Alice: Would you tell me, please, which way I ought to go from here?
The Cat: That depends a good deal on where you want to go.
Alice: I don't much care where.
The Cat: Then it doesn't matter which way you go.
Alice: ... so long as I get somewhere.
The Cat: Oh, you're sure to do that, if you only walk long enough."

Lewis Carroll: Alice in Wonderland

One could initially ask whether the present overall strategic allocation of company resources to different functional areas is something sacred. Is it possible that money spent on innovation should be allocated to marketing, or that money spent on production should be allocated to innovation, depending on changes in the external business environment? The answer may be different if this discussion is carried out at top management level, among the R&D management team, or on the company production plant floor. One must regrettably say that regardless of forum and corporate level, the answers are probably equally unreliable, since they are often snapshots of the present, invariably mistaken for images of the future. The interesting point is, however, not that a good answer is very difficult to find, but the amazing thing is that the answer is sometimes never even sought.

Even if a company lacks any kind of explicitly stated innovation strategy there is nevertheless an untold, unspoken strategy, which in this book will be conceptualized and called the "hidden strategy". The allocation of total resources to innovation is then the mirror of such an overall hidden innovation strategy. It is reflecting the importance a company's management attaches to innovation as a contributor to future business performance, whether explicitly stated or not. The further allocation of those resources to different kinds of R&D activities (the distributed innovation intensities), and the distribution of available resources to process innovation, is then the mirror of the hidden overall process innovation strategy. Following this line of thought, the further distribution of resources allocated to process innovation among different kinds of activities and the "newness" of the associated portfolio of process innovation projects are then the mirror of this more detailed hidden process innovation strategy. Such a strategy, regrettably, is sometimes not even articulated at all or further communicated to internal stakeholders for process innovation.

This chapter will start with an introduction of strategy making and roadmapping (TRM), followed by a discussion of the concept of the "hidden" innovation strategy. Finally, the development of a Strategic Research Agenda (SRA) for process innovation will be presented.

6.1 Technology roadmapping — translating business strategy into innovation objectives

A brief introduction to strategy and strategy-making

Strategy can be derived from an old concept in warfare, and part of one of the classics by Clausewitz is still enjoyable reading for managers (von Clausewitz, 1984). Strategy as a management concept is however fairly young, and Drucker (1985) regretted in the early sixties that he was not one of the first users of such a concept. An early presentation of strategy as a management tool was instead made by

Ansoff (1965). Drucker (1993) discusses such management activities as "action patterns" and continues:

> "These are the assumptions that shape any organization's behaviour, dictate its decisions about what to do and what not to do, and define what the organization considers meaningful results. These assumptions are about markets. They are about identifying customers and competitors, their values and behaviour. They are about technology and its dynamics, about company strengths and weaknesses. These assumptions are about what a company gets paid for. They are what I call a company's theory of the business."

What Drucker yesterday called the "theory of business" (Chapter 1, Drucker, 1998) is today more considered as and often referred to as a firm's business model (Johnson *et al.*, 2008). While Drucker focuses more on different kinds of "hands-on" strategies, easily usable even for strategists in small companies, one dominant strategy scholar — Mintzberg — also discusses the strategy-making process in depth (Mintzberg, 1987):

> "In my metaphor, managers are craftsmen and strategy is their clay. Like the potter, they sit between a past of corporate capabilities and a future of market opportunities. And if they are truly craftsmen, they bring to their work an intimate knowledge of the material at hand. That is the essence of strategy. ...
>
> My thesis is simple: the crafting image better captures the process by which effective strategies come to be. The planning image, long popular in literature, distorts these processes and thereby misguides organisations that embrace it unreservedly."

This picture of strategy-making probably also appeals to strategy-makers in an R&D organization, because it depicts strategy-making as a creative exercise which cannot be performed without an intimate knowledge of the subject area of innovation. The following strategic concepts created by Mintzberg are also useful for every strategy-maker, whether creating a corporate strategy, company strategy, innovation strategy or process innovation strategy (Mintzberg and Waters, 1985):

- Intended: a strategy which appears worth pursuing.
- Deliberate: a strategy which has advanced from "intended" and become a purposeful and planned strategy.
- Unrealized: an "intended strategy" which for some reason was never implemented.
- Emergent: a strategy driven by external forces that just happened.
- Realized: a deliberate or emergent strategy that has been implemented.

Mintzberg thus emphasizes the dynamic aspects of strategy-making; in later publications he also strongly opposes the need for a company strategy-planning department (Mintzberg, 1994). The Mintzberg view of strategy-making is well captured in the following quotation:

> "The popular view sees the strategist as a planner or as a visionary, sitting on a pedestal dictating brilliant strategies for everyone else to implement. ... I wish to propose an additional view of the strategist as a pattern-recogniser — a learner if you will — who manages a process in which strategies (and visions) can emerge as well as being deliberately conceived. I also wish to redefine the strategist, to the extent that somebody in the collective entity made up of the many actors whose interplay speaks an organisation's mind. This strategist finds strategies no less than creates them, often in patterns that inadvertently form their own behaviours."

The competitive perspective on strategies was introduced by Porter (1980), and although Drucker's strategic types could sometimes be more useful for more company-specific applications, Porter's three Generic Strategies are still necessary and very useful concepts in the corporate world:

- Special market: Focus on a particular category of buyer, segment of the product range or geographical market. Whereas the overall cost leadership and differentiation strategies aim at achieving their objectives in the industry at large, the whole special market strategy aims primarily at selling to a specific target group, and every departmental programme is designed accordingly.

- Differentiation: The second basic strategy implies differentiating the product or service the company offers and thereby creating something which is perceived as unique of its kind. Differentiation can take many forms: design and brand profile, technology, refinements, customer service, dealer network or other dimensions. It is best if a company can differentiate itself in several dimensions. ... It must be emphasised that the differentiation strategy does not mean that the company can disregard costs though they are not the strategic aim. ... Differentiation may sometimes make it impossible to capture a large market share; it often calls for a certain degree of exclusiveness.

- Overall cost leadership: Calls for a massive effort to create optimised production facilities, tight control of operating costs and general overheads, avoidance of marginal customers, and cost minimisation of areas such as R&D, service and advertising. ... Low cost compared to competitors is the all-pervading theme of this strategy, though quality, service and other areas cannot be neglected. ... Achieving cost leadership often requires a relatively high market share or other advantages such as good access to raw materials.

Porter's three so-called Generic Strategies give food for thought to makers of innovation strategies. The overall cost leadership strategy emphasizes the need to keep R&D costs low and to secure a strong raw material supply. Low costs for product development are then likely to work well, but perhaps process innovation could be more strongly articulated as one important tool for keeping production costs low. If we consider the discussions in Chapter 4 together with Porter's view, the relationship between product innovation and process innovation might then be an important aspect. On the other hand, the differentiation strategy applied in the process industries should invite a strong focus not only on product innovation but on application development and customer support as well.

This very brief and rather incomplete introduction to the discipline of strategy and strategy-making is however simply intended to emphasize the importance of the development of a TRM well founded in the company's overall business strategy. Such a roadmap will serve as a guideline for the development of a sound SRA and a process innovation strategy in particular.

Technology roadmapping: a framework for innovation strategy

Roadmappping is increasingly being used in different applications since its first introduction in the late 1980s. In support of product-technology planning it is foreseen that roadmapping will be increasingly used as a core integrating mechanism for supporting strategic dialogue (Phaal *et al.*, 2008). Generally, a TRM comprises a number of "layers" on a time-based chart; technology programmes are linked to future products and services and to market and business opportunities (Farrukh *et al.*, 2003). Roadmapping can thus be looked upon as a planning tool and as such it is now often also used on a national or industry level. In the EU, for example, the research allocations for all industrial "framework programmes" must be founded on sectorial technology platforms including long-term (10–15 year) TRMs. Those technology platforms and roadmaps are created by leading European industrial corporations within each sector and thus give an overall agenda for the creation of multiple strategic research agendas within the EU.

On a corporate level, the roadmapping tool is used for many applications from strategic business to others which are more product and technology specific. Technology roadmapping started early at Motorola (Richey, 2004), and is forcefully pursued at Philips (van Doorn, 2006). In the integration of R&D management and the business management strategy Six Sigma, the use of TRM and QFD is presented by Park and Gil (2006) in their paper about how Samsung transformed its corporate R&D centre. Samsung is rapidly leapfrogging from technology follower to global leader, and in this process their large corporate R&D centre is building close relationships with business divisions by sharing business strategies and technology roadmaps.

Corporate TRMs are often product-related, and can be defined and presented in many ways; the following definition by Garcia and Bray (1997) is one alternative:

"Technology roadmapping is a needs-driven technology planning process to help identify, select, and develop technology alternatives to satisfy a set of product needs. ... Given a set of needs, the technology roadmapping process provides a way to develop, organise, and present information about the critical system requirements and performance targets that must be satisfied by certain time frames. It also identifies technologies that need to be developed to meet those targets."

When the product TRM is used as a platform for future R&D, the means to reach the targets by development work is often called a Strategic Research Agenda (SRA). In the development of such a research agenda, there is a need to relate those targets to necessary corporate R&D in many different areas. Whalen (2007) writes:

"For example, a manager or product owner of a specific product line should own the product platform roadmap. ... In general, the map should contain information on the current products/platforms available on the market and at least two generations of new products evolving from that platform. ... Specific key attributes/specifications for each product that are required for its financial success (validated with marketing). These attributes should represent those key specifications that technology/capability development will target."

The usefulness of TRMs as instruments for the creation of strategic research agendas is today fully recognized in industry and academia. How to create such roadmaps in practice is, however, something that needs more guidance and development.

In an action-based research project by Farrukh *et al.* (2003), the dimensions selected for the roadmap were market, product and technology. Afterwards, possible technical solutions were identified that could deliver the desired product features. The authors concluded that this approach was similar to QFD! After a product, or product category, has been selected for the creation of a TRM, the most critical first activity is to identify and agree upon what product needs must be satisfied. In other words, future targets for an SRA must be set and related process technology identified.

The further development of such a roadmap is seldom too difficult, but the identification of future fact-based product needs and targets is a crucial platform upon which the quality of all other work rests. But how do we identify those future targets without relying on guesswork or old company market and production conceptions, and how do we go from art to science in the development of an SRA to meet such targets? In their paper on customized roadmapping, Phaal *et al.* (2004) argue that the capabilities of the roadmapping approach must be matched to the business issues being addressed.

In the discussion of the next generation of roadmapping for innovation planning, Phaal *et al.* (2008) focus on the "fuzzy front end" of innovation and state that the innovation roadmapping method includes end-user insight generation and proposition validation, concept creation and validation, product planning, technology planning and competence planning. In creating a multifunctional roadmap, trends in product-related customer requirements are shown, as well as the way target specifications for these products are derived from those trends in customer requirements. It was concluded that further work is needed to understand what constitutes a core set of management tools and frameworks that can address strategic planning and innovation challenges, and how they fit together.

Roadmapping products, process technology and raw materials in the process industries

Because of the integration of different innovation activities in the process industries it is not enough to draw a roadmap for process innovation; there is a need for three well integrated roadmaps for products, process technology capabilities and raw materials.

Product roadmap

A product-TRM is a structured plan for company short-term incremental product improvements (product care), as well as a long-term plan for product renewal. This roadmap must align well with the company's overall business strategy. It must also integrate and bridge product improvements and the transition into new product generations.

The complexity of making a good product-TRM depends on the necessary company integration of valid product/market scenarios and assessment of future necessary production capabilities, including future specifications of raw materials. In short:

- A good product roadmap is the art and science of figuring out how a company's long-term product portfolio should develop and change.
- Product roadmapping is about phasing out old products and phasing in new ones.
- Product roadmapping is deciding how existing products must be improved or can be allowed to deteriorate.
- Product roadmapping is about understanding future market needs and potential technological improvements to give guidance for product innovation and striking a proper balance between market pull and technological push.

Process technology roadmap

Technology planning, previously presented in Chapter 4, can serve as part of an inventory, an analysis and an important input to a corporate roadmap for the acquisition and phase-out of corporate process technology. A good and well-realised process-TRM can play an important part in determining a company's future competitiveness in its sector of the process industries. The starting-point for process technology roadmapping is the company's present process technology and the overall context is the company's business strategy and product portfolio. It must however be observed that part of the input for the process-TRM comes from product development needs, but the main part of the process-TRM comes from pure production process technology needs for the innovation of more cost-efficient production technology. In a similar vein, part of the input for the process-TRM should come from the raw material roadmap securing the development of the raw material supply base.

Raw material roadmap

There is also a need to formulate long-term objectives and a roadmap for corporate raw material supplies. Price increases and large fluctuations in raw material prices tend to make many firms in the process industries too sensitive and dependent on the cost of raw materials. The importance of such a raw material-TRM is then illustrated and easily estimated by the calculation of the raw material share of corporate total production costs. Since raw materials however are sometimes not interchangeable in the company's production system, they cannot be considered as "pure commodities" but more bordering on functional products. More price-competitive alternative sources for raw material supplies, in combination with innovative process and product solutions matching those alternative raw materials, can create interesting opportunities for profitable new solutions to be sought and incorporated in such a raw material-TRM.

Developing three integrated roadmaps

Product, process technology and raw material roadmaps can be outlined using corporate business strategy as a backbone. A symbolic structure for such a roadmap is presented in Figure 6.1.

Creating a strategic research agenda following the technology roadmaps

When corporate innovation strategies are not explicitly well developed, or if the more detailed strategies for process innovation, application development, etc., are non-existent, the conclusion may then perhaps be that the company does not have such strategies. It is

Roadmaps for the Process Industries	Five-year perspective (1-5 years)	Ten-year perspective (5-10 years)	Fifteen-year perspective (10-15 years)
A product roadmap (including development targets and product specifications)	Product generation A		
		Product generation B	
			Product generation C
A process technology roadmap (including development targets and process capabilities)	Incrementally improved process technology type A		
	Incrementally improved process technology type B		
		Radical new process technology type A	
A raw material roadmap (development targets for volumes and quality)	External raw materials supplied		
		Captive raw material supply	

Figure 6.1   A symbolic and simplified roadmap for products, process technology and raw materials, to be used in the creation of an SRA for process innovation.

however argued in this chapter that such strategies do exist, but since they not are explicitly formulated they are called hidden strategies. Though it lacks a well formulated strategy, the company nevertheless has an "action pattern" (see the earlier reference to Drucker's work), which will be discussed in the following section.

In the strategy-making process it is of vital importance never to forget that R&D is a means, not the final objective. The development of TRMs will thus ensure that corporate overall business objectives guide R&D, and not vice versa.

However, the incentive for R&D to present new findings that may change overall business strategy must always be encouraged; we can refer to this interaction in traditional terms as a balancing of "market pull" and "technology push", (see Figure 6.2).

It is further proposed that in the creation of future roadmaps there is a need for:

- An improved process for generating future product specifications and customer-related target values for the selected products or product categories.

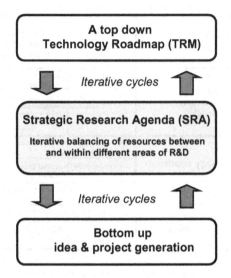

Figure 6.2   A conceptual model of the interactive strategy making process.

- An improved process for generating targets for process capabilities related to future product specifications and target formulation and targets generated by production in order to secure more cost-efficient production technology.
- An improved process for generating future raw material specifications and needs for development of captive raw material supplies.

It is argued that the creation of such roadmaps, and the further development of associated research agendas, can be greatly facilitated in the process industries by the use of the mpQFD system.

6.2 A strategic research agenda translated into distributed innovation intensities and "newness"

In the process industries a wide spectrum of different innovation and innovation-related R&D must be addressed in future SRAs (Lager, 2008).

Firms that have created technology roadmaps strongly related to overall business strategies and have a well-functioning iterative creation of a SRA are in a good position and can very likely look forward to profitable future development. The deliberate strategies agreed upon should then be translated into necessary monetary resources that must be allocated to R&D (R&D intensity). Then the total resources for R&D should also be translated into different areas for R&D (distributed innovation intensities) as well as a characterization of allocated resources to innovation of different degrees of newness.

Innovation success in the process industries depends thus on the creation of a balanced innovation strategy. The importance of raw material development with suppliers and/or securing a competitive supply of captive raw materials must not be ignored. Applied research and product and process innovation activities must also strike a good balance. Innovation-related activities, which are often organized and carried out by R&D, supporting or complementing the innovation activities, must also get their proper share of corporate resources and be well integrated in an overall holistic approach to innovation.

However, companies regrettably sometimes lack such overall functional work processes for strategy-making or use work processes that are less efficient and effective. Even if overall allocation of innovation resources is based on a return-on-investment model and a selection made on a project-by-project base, such an allocation procedure is not currently considered appropriate for an overall effective innovation strategy development process. Lacking better models for resource allocation to innovation, the "meta rule" model of allocating resources to R&D at fixed percentages is then often not only superior but also more commonly used. But the two questions still remain: How much is enough? How to spend it? (Lager and Blanco, 2010).

We all agree that corporate management in a market economy is about economizing scarce resources. It becomes more difficult, however, when we must decide how those scarce resources ought to be distributed and, more specifically, how much of corporate resources should be allocated to R&D. It is often generally agreed that the company's innovation strategy must be well aligned within its overall business strategy. If for example the company has opted to compete as a commodity producer, and chosen a strategic direction we can call "competing by cost", the innovation strategy should of course support such a strategic choice (Porter, 1980), favouring a strong process innovation strategy. Innovation strategy-making is discussed here on the lowest hierarchic level in a decentralized corporation. However, in the last part of the book we will further consider how such decentralized innovation strategies can and should be aggregated to a higher corporate group level.

How much innovation is enough?

This may seem a trivial question, but relevant and more fact-based resource allocation models for innovation are required to answer it. In the general corporate resource allocation process, the distribution of corporate total resources to different functional areas like sales & marketing, production and R&D is often, as stated before, based on traditional percentages (meta rules) and yesterday's needs (Christensen and Raynor, 2003). But how can we better determine the appropriate yearly investment in innovation within the overall company budget and strategic investment plans? A strategy-making process based on well-developed TRMs and associated SRAs is naturally the recommendation in this book.

In this book we take a rather pragmatic view of companies' intended strategy from a money distribution perspective; money distributed to different "buckets" of the firm's organization (Cooper *et al.*, 1997). The underlying philosophy is that the pattern of how available resources are spent on innovation is the image of the firm's intended deliberate strategy, developed according to Fig. 6.2, or its unspoken "hidden" strategy for innovation.

In the general corporate resource allocation process, the distribution of total resources to different functional areas is however often based on traditional percentages. This applies in particular to the general procedure for the distribution of corporate resources to innovation. Such traditional "industrial pseudo-models" for distribution of resources to innovation may be:

- R&D will get as much as we can afford this year.
- R&D will get as much as they usually get each year.
- R&D will get an average for our industry sector.

It is rather surprising that the author's anecdotal industrial experience suggests that corporate excellence in this area seems to be high on a project level, but gradually diminishes with increasing level in the organization. Unfortunately this notion is confirmed in academic research (Bower and Gilbert, 2005). How can resources be better allocated to innovation in a short-term, medium-term and long-term perspective? The topic of resource allocation is surprisingly little researched, but there are a few references (Bower, 1970; Bower and Gilbert, 2005; Cooper and Kleinschmidt, 1988; Hamilton, 2006). One is seriously inclined to believe that current practices are preventing a sound and strategic resource allocation to innovation for the future competitive and profitable corporation!

How to spend it?

Resource allocation models for the distribution of innovation resources to different areas of innovation or innovation-related activities are totally lacking. How then to determine the appropriate yearly investments for different areas of innovation within the overall company R&D budget and strategic investment plans? Once again, TRMs and associated SRAs are recommended as excellent tools for such decisions. Even so, resource allocation to R&D in general and to different areas of R&D should be better and more explicitly presented in the future. This is because in the future the process industries must pay more attention to finding a balanced distribution between a wide spectrum of innovation and innovation-related activities. The typology and structure for such different areas of innovation and

Activities at the R&D department			How much is allocated to the area today (%)	How much should be allocated to the area in the future (%)
Development of captive raw material supply				
Development with raw material suppliers				
Innovation with the internal customer		Process development		
		Industrialization		
		Internal technical support		
Innovation with the external customer		Product development		
		Application development		
		External customer support		
Applied research				
Basic research				

Figure 6.3   Distributed innovation intensities in the process industries.

innovation-related activities presented in Part 1 are listed and used for such a presentation in Figure 6.3.

In many companies some of these areas have presumably not yet been identified or recognized as important areas in the internal R&D resource allocation process. Following different corporate roadmaps and strategic routes for R&D, the importance and relevance of the listed R&D areas will probably differ between different sectors of the process industries and between firms within each sector. Furthermore the different R&D areas will very likely differ between different business segments, divisions and subsidiary companies within a large multinational corporation.

The above more structured view of different areas of R&D is not however intended to imply a stronger separation of the activities. On the contrary, a better integration and dynamic collaboration is desirable.

Discover and challenge the "hidden innovation strategy"

Resources allocated to R&D should reflect the overall importance of R&D to the company. In other words, R&D spending ought also to be strongly related to the overall corporate business strategy. For each organizational unit in a corporation, the previously presented innovation intensity is thus a mirror of the "hidden" innovation strategy. In all corporations, each organizational unit should present not only its innovation intensity (and absolute figures on innovation resource allocation) but also trend curves on how those figures have changed during the past 10 years. But not only that, how are those figures expected to change over the next five years? Those figures should then be commented upon and aggregated for the total corporation. Since some of those figures are generally also available for competitors, a comparison could provide food for thought.

6.3 A strategic research agenda for process innovation

The following colourful metaphor (Mintzberg, 1994, p. 366) illustrates the creative nature of strategy-making. Mintzberg's expresses concepts of deliberate and emerging strategies and the dynamic nature of the strategy process:

> "To return to our grass roots model of strategy making, the planners have to be looking for growths that, while seeming like weeds, are in fact capable of bearing fruit. These have to be watched carefully, sometimes without disturbing their emergent nature until their worth has become clear. Only then can they be propagated as valuable plants, or else cut down as worthless."

The innovation strategy making process must also in the future be more formalized as a work process and be related better to the firm contextual environment (Larsson, 2006; Larsson and Bergfors, 2006; Larsson and Sugasawa, 2007).

Strategic process innovation

Even excellent companies in the process industries must continually improve their performance. One way to achieve a competitive advantage is to use process and product development to improve the cost-efficiency of production processes and the performance of products. In the process industries, where investments in production technology are often very costly and sometimes risky, it is argued that in many industry sectors long-term strategy and planning of process innovation is vital — sometimes even more important than product innovation — because:

- Investments in new or improved production technology will influence future production flexibility and largely determine long-term production costs.
- Present process technology will limit possibilities for product differentiation and will always influence product quality.
- New investments in a company's process technology will sometimes determine its ability to produce new products and product qualities 10–15 years ahead.

The selection and development of company process technology is thus of high importance, often more important to companies in the process industries than in other manufacturing industries.

Delineate a winning process innovation strategy

An allegory to Robert Cooper's (1993) title of his book "Winning at new products" could now be "Winning at new processes". Process innovation is an activity with many facets, and it is impossible to create a solid process innovation strategy as a solitary activity. The message is that the development of a process innovation strategy must be integrated in the overall innovation strategy-making process, in the iterative manner presented in Figure 6.4.

By referring to Figure 6.2 and using the TRMs, associated SRAs can be developed using, among many other sources of information, the wealth of information available from the different QFD matrices presented in Chapter 5. Figure 6.5 shows a symbolic SRA for process innovation and innovation-related activities.

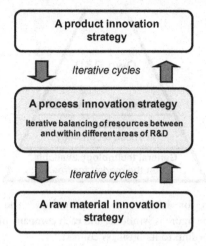

Figure 6.4   The iterative development of integrated SRA in the process industries.

A Strategic Research Agenda for process innovation		Five-year perspective (1–5 years)	Ten-year perspective (5–10 years)	Fifteen-year perspective (10–15 years)
Product development related development of process technology		Project portfolio	Project portfolio	
Production related	**Process innovation**		Project portfolio / Project portfolio	
	Development with equipment suppliers		Project portfolio	
	Internal technical support			
	Industrialization		Project portfolio	
Raw material related development of process technology			Project portfolio	

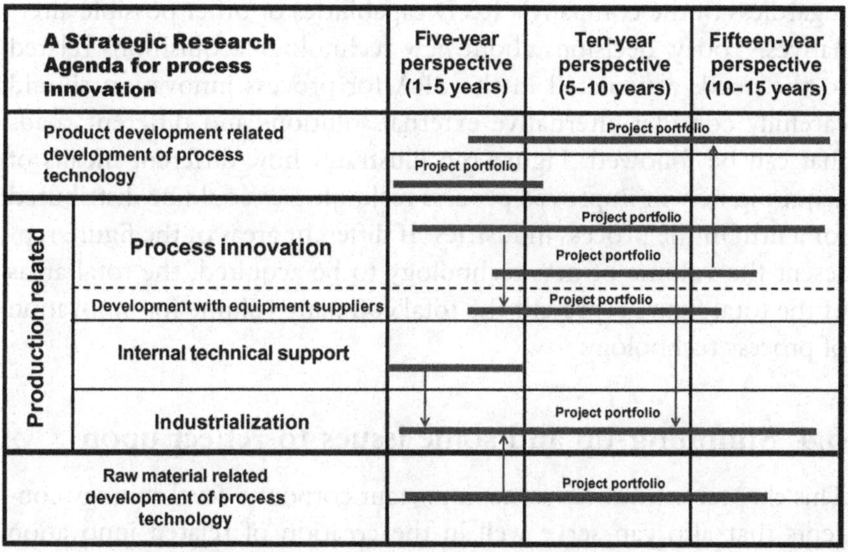

Figure 6.5   A symbolic and simplified SRA for process innovation and innovation-related activities for internal customers.

Strategic technology acquisitions

In the old days, acquiring new company process technology in the process industries was often synonymous with in-house R&D,

Figure 6.6 Options for acquiring new technology. The relative distribution between the areas in the figure is symbolic, and each company must seek its own optimum distribution according to its business context.

regardless of the company's R&D capabilities or other possible alternatives. Today, decisions about new technology acquisitions related to the needs articulated in the SRA for process innovation should carefully consider alternative external solutions and different roads that can be followed. Figure 6.6 illustrates how different means of acquiring new or improved process technologies could be distributed for a firm in the process industries. If different areas of the figure represent the volume of new technology to be acquired, the total areas of the total figure represent the total company volume for innovation of process technology.

6.4 Summing-up and some issues to reflect upon

This chapter introduces some important corporate-level strategy concepts that also can serve well in the creation of related innovation strategies. The message is that different business strategies call for different innovation strategies, and tomorrow's competitive corporate environment needs better guidelines and models for allocation and distribution of available corporate resources to innovation.

The danger of having a "hidden company strategy" and an associated "hidden innovation strategy" is that innovation activities can take directions dictated more by individual desires and motives, and that resources for innovation may be allocated to more short-term needs. For the R&D organization, and process innovation in particular, internal allocation of resources to technical support instead of to long-term strategic process innovation is thus not uncommon. But if no innovation strategic intent is articulated, any kind of resource allocation is naturally accepted, see further the introductory quotation for this chapter.

The development of a SRA is discussed, and the important interaction between this agenda and the TRM is stressed. It is further pointed out that the interdependence between raw material properties, process technology and the performance of finished products makes a system-related approach to the development of innovation strategies important in the process industries. In the further development of a process innovation strategy, the interaction between the SRAs for products, process technology and raw materials is thus highlighted.

- *Does your company have a strategy and a strategy process? If so, what does your strategy development process look like, and how transparent is this strategy at all levels in the company?*
- *Referring to Porter's three generic strategies, which one do you find pictures your company's explicit or "hidden" strategy the best?*
- *In your strategy and strategy-making, does your company have an explicit innovation strategy process?*
- *If your company has an innovation strategy, how well is it connected to overall business strategy, and how well is it communicated to its stakeholders?*
- *What is your company's innovation intensity, and what underlying information and considerations is this "hidden" overall innovation strategy based upon?*
- *Using your company's innovation intensity as a point of departure, how are the total innovation resources deployed, and more specifically, to which innovation and innovation-related areas presented in this chapter are they allocated? What are the company's motives for this selected resource allocation?*

- *Does your company have an explicit process innovation strategy, or do you think that long-term process innovation does not matter so much?*

References

Akao, Y (ed.) (1990). *Quality Function Deployment: Integrating Customer Requirements into Product Design*. Cambridge, MA: Productivity Press.

Akao, Y (2003). The leading edge in QFD: past, present and future. *International Journal of Quality & Reliability Management*, 20, 20–35.

American Supplier Institute (1989). Quality Function Deployment — Three Day Workshop. Dearborn, American Supplier Institute Inc.

Anderson, P and Tushman, ML (1990). Technological discontinuities and dominant designs: a cyclical model of technological change. *Administrative Science Quarterly*, 35, 604–633.

Ansoff, HI (1965). *Corporate Strategy*. New York: McGraw-Hill.

Benner, M, Linnemann, AR, Jongen, WMF and Folstar, P (2003). Quality Function Deployment (QFD) — can it be used to develop food products? *Food Quality and Preference*, 327–339.

Booz, Allen and Hamilton (1982). *New Products for Management for the 1980s*. New York: Booz, Allen & Hamilton.

Bower JL (1970). *Managing the Resource Allocation Process*. Boston, MA: Harvard Business School Press.

Bower, DJ and Keogh, W (1996). Changing patterns of innovation in a process-dominated industry. *International Journal of Technology Management*, 12, 209–220.

Bower, JL and Gilbert, C (eds.) (2005). *From Resource Allocation to Strategy*. Oxford: Oxford University Press.

Carroll, L (2000). *Alice's Adventures in Wonderland and Through the Looking Glass*. New York: Signet Classics.

Christensen, CM and Raynor, ME (2003). Why hard-nosed executives should care about management theory. *Harvard Business Review*, September, 67–74.

Cobbenhagen, J, den hertog, F and Philips, G (eds.) (1990). Management of Innovation in the Processing Industry: A Theoretical Framework. In *New Explanations in the Economics of Technology and Change*, C Freeman and L Soete (eds.), pp. 55–73. London, Pinter Publishers.

Cooper, RG (1993). *Winning At New Products. Accelerating the Process from Idea to Launch*. Addison-Wesley Publishing Company Inc.

Cooper, RG and Kleinschmidt, EJ (1988). Resource allocation in the new product process. *Industrial Marketing Management*, 17, 249–262.

Cooper, RG, Edgett, SJ and Kleinschmidt, EJ (1997). *Portfolio Management for New Products*, Hamilton, Ontario: McMaster University.

De Oliveira, CA, Magri, W and Torres, IV (1996). QFD in a Brazilian Steel Company. *Transactions from The Eighth Symposium on Quality Function Deployment*. Novi, MI: QFD Institute.

Drucker, P (1985). *Innovation and Entrepreneurship*.

Drucker, P (1998). *On the Profession of Management*. Boston, MA: Harvard Business School Press.

Dussauge, P, Hart, S and Ramanantsoa, B (1987). *Strategic Technology Management*. New York: John Wiley & Sons.

Farrukh, C, Phaal, R and Probert, D (2003). Technology roadmapping: linking technology resources into business planning. *Int. J. Technology Management*, 26, 2–19.

Foster, RN (1986). *Innovation — the Attacker's Advantage*. New York: Summit Books.

Freeman, C (1990). Technical innovation in the world chemical industry and changes of techno-economic paradigm. In *New Explorations in the Economics of Technological Change*, C Freeman and L Soete (eds.), pp. 74–91. London: Pinter Publishers.

Garcia, ML and Bray, OH (1997). Fundamentals of Technology Roadmapping. Albuquerque NM: Sandia National Laboratories.

Ginn, D and Zairi, M (2005). Best practice QFD application: an internal/external benchmarking approach based on Ford Motors' experience. *International Journal of Quality & Reliability Management*, 22, 38–58.

Granger, RJ Value for money in R&D. Arthur D. Little International Inc.

Griffin, A and Page, AL (1991). PDMA Success measurement project: Recommended measures for product development success and failure. *Journal of Product Innovation Management*, 13, 478–496.

Hamel, G and Heene, A (1994). *Competence Based Competition*. Chichester: John Wiley & Sons.

Hamilton, RW (2006). When the means justify the ends: Effects of observability on the procedural fairness and distributive fairness of resource allocation. *Journal of Behavioural Decision Making*, 19, 303–320.

Hanson, D (1993). Quality Function Deployment for product and service improvement. *Transactions from the Fifth Symposium on Quality Function Deployment*, pp. 363–377. Novi, MI: QFD Institute.

Hauser, JR and Clausing, D (1988). The house of quality. *Harvard Business Review*, May–June, 63–73.

Hayes, RH, Pisano, GP and Upton, DM (1996). *Strategic Operations: Competing Through Capabilities.* New York: The Free Press.

Herrman, A, Huber, F, Algesheime, R and Tomczak, T (2006). An empirical study of quality function deployment on company performance. *International Journal of Quality & Reliability Management*, 23, 345–366.

Hoque, M, Akter, M, Yamada, S and Monden, Y (2000). How QFD and VE should be combined for achieving quality & cost in product development. In *Japanese Cost Management*, Y Monden (ed.), London: Imperial College Press.

Hutcheson, P, Pearson, AW and Ball, DF (1995). Innovation in process plant: a case study of ethylene. *Journal of Product Innovation Management*, 12, 415–430.

Jiang, J-C, Shiu, M-L and Tu, M-H (2007). QFD's evolution in Japan and the West. *Quality Progress*, 40(7), 30.

King, B (1987). *Better Design in Half the Time — Implementing QFD Quality Function Deployment in America.* Methuen, MI: GOAL/QPC.

Kubota, I (1990). Using Quality Function Deployment: The case of Nippon Carbon. In *Quality Function Deployment Integrating Customer Req-uirements into Product Design*, Y Akao (ed.), Cambridge MA: Productivity Press.

Lager, T (2002a). Product and process development intensity in Process Industry: A conceptual and empirical analysis of the allocation of company resources for the development of process technology. *International Journal of Innovation Management*, 4, 105–130.

Lager, T (2002b). A structural analysis of process development in process industry — A new classification system for strategic project selection and portfolio balancing. *R&D Management*, 32, 87–95.

Lager, T. (2005a). The industrial usability of quality function deployment: a literature review and synthesis on a meta-level. *R&D Management*, 35, 409–426.

Lager, T (2005b). Multiple Progression — a proposed new system for the application of Quality Function Deployment in process industry. *International Journal of Innovation Management*, 9, 311–341.

Lager, T (2008). Using Multiple Progression QFD for roadmapping product and process related R&D in the process industries. *14th International Symposium on Quality Function Deployment*, Beijing, China.

Lager, T and Kjell, Å (2007). Multiple Progression QFD: a case study of cooking product functionality at Arla Foods. In *13th International & 19th*

North American Symposium On QFD, G. Mazur (ed.), pp. 271–295. Williamsburg, VA: QFD Institute.

Linn, RA (1984). Product development in the chemical industry: a description of a maturing business. *Journal of Product Innovation Management*, 2, 116–128.

Martins, A and Aspinwall, EM (2001). Quality function deployment: an empirical study in the UK. *Total Quality Management*, 12, 575–588.

Mintzberg, H (1987). Crafting strategy. *Harvard Business Review*, July–August, 66–75.

Mintzberg, H (1994). *The Rise and Fall of Strategic Planning*. London: Prentice Hall Europe.

Mintzberg, H and Waters, JA (1985). Of strategies, deliberate and emergent. *Strategic Management Journal*, 6, 257–272.

Miyashita, K and Russel, D (1995). *Keiretsu: Inside the Hidden Japanese Conglomerates*. New York: McGraw-Hill.

Mizuno, S and Akao, Y (eds.) (1994). *QFD: The Customer-Driven Approach to Quality Planning and Deployment*. Tokyo: Asian Productivity Organization.

Ootaki, A, Oyaizu, M and Koura, K (1996). How to connect material and technology seeds to customer needs. *Transactions from the Eighth Symposium on Quality Function Deployment*, pp. 243–259. Novi, MI: QFD Institute.

Park, S and Gil, Y (2006). How Samsung transformed its corporate R&D center. *Research Technology Management*, 24–29.

Phaal, R, Farrukh, C and Probert, D (2004). Customizing roadmapping. *Research Technology Management*, 47, 26–37.

Phaal, R, Simonse, L and Den Ouden, E (2008). Next generation roadmapping for innovation planning. *Int. J. Technology Intelligence and Planning*, 4, 135–152.

Pisano, GP (1997). *The Development Factory: Unlocking the Potential of Process Innovation*. Boston, MA: Harvard Business School.

Porter, ME (1980). *Competitive Strategy: Techniques for Analyzing Industries and Competitors*. New York: Free Press.

Pugh, S (1981). Concept Selection — A method that works. *International Conference on Engineering Design*, pp. 497–506. Rome, Italy.

Rainbird, M (2004). Demand and supply chains: the value catalyst. *International Journal of Physical Distribution & Logistics Management*, 34, 230–250.

Richey, JM (2004). Evolution of roadmapping at motorola. *Research Technology Management*, 45, 37–59.

Scheurell, DM (1992). Taking QFD through to the production planning matrix: putting the customer on the line. *Transactions from the Fourth Symposium on Quality Function Deployment*, pp. 532–543. Novi, MI: QFD Institute.

Scheurell, DM (1994). Beyond the QFD house of quality: using the downstream matrices. *World Class Design in Manufacture*, 1, 13–20.

Skinner, W (1992). The shareholder's delight: companies that achieve competitive advantage from process innovation. *International Journal of Technology Management*, 41–48.

Stitt, J and York, C (1993). Just do it. *Transactions from the Fifth Symposium on Quality Function Deployment*, pp. 419–440. Novi, MI: QFD Institute.

Utterback, JM (1994). *Mastering the Dynamics of Innovation: How Companies Can Seize Opportunities in the Face of Technological Change*. Boston, MA: Harvard Business School Press.

Utterback, JM and Abernathy, WJ (1975). A dynamic model of process and product innovation. *Omega* 3, 639–655.

Walters, D (2006). Demand chain effectiveness-supply chain efficiencies. *Journal of Enterprise Information Management*, 19, 246–261.

Van Doorn, M (2006). EMC technology roadmapping: a long-term strategy. Eindhoven, The Netherlands, March 2006.

Whalen, PJ (2007). Strategic and technology planning on a roadmapping foundation. *Research Technology Management*, 50, 40–51.

Von Clausewitz, C (1984). *On war*. 1976 reprint, Princeton: Princeton University Press.

Zheng, LY and Chin, KS (2005). QFD based optimal process quality planning. *International Journal of Advanced Manufacturing Technology*, 831–841.

PART 3

LEAN PROCESS INNOVATION

In Part 2 we discussed the development of a process innovation strategy was discussed. Under the banner of "lean process innovation", this part of the book will consider various means of implementing such a strategy with a view to achieving better R&D performance. Organizational matters are usually high on the company's agenda for achieving good company and R&D performance. The advantages of developing innovative and well-functioning corporate organizational solutions are that they create a strong competitive edge, because they are often difficult to copy. An R&D organization set up to support efficient and effective corporate innovation is thus not only a valuable and intangible asset, but also essential to well-functioning organizational learning.

The development of process technology will be one important area in the future not only for commodity producers, but for producers of more functional products in the process industries (Fig. 2.3). In case studies from the pharmaceutical industry, Pisano (1997, p. 9) argues for an expanded view of the role of process development, one that recognizes how process development capabilities can be valuable even outside the confines of so called mature, cost-driven industries.

The "lean" concept will be used in the same sense as in production processes, where lean production focuses on more efficient resource utilization and eliminating factors that do not create value for the end user (Liker and Meier, 2006). In a similar vein, better organizational structures and better functioning work processes for process innovation aim at creating more value for the firm and its customers for less input of work.

The use of a more efficient work process could also be an important road to follow to improve future process development performance, even for firms that develop and use emerging technologies like bio- and nano-technologies. We must however regrettably observe that the implementation of company innovation work processes, except for the product development process, is still in its infancy in the process industry (Lager, 2000). This area will therefore be dealt with at greater length in this part of the book.

The development of formal work processes for process innovation is so far a white space on the map of innovation work processes. They are therefore presented in depth in Chapters 8 and 9. One of the most important aspects of such work processes is to stimulate and improve organizational learning. This, together with more efficient work processes, is an important objective, with lean process innovation as an outcome.

Chapter 7

Designing a Structural Organization for Process Innovation

"The goal should be to identify the organisational configurations most suited to specific technological and market environments, rather than to seek a single ideal or best-practice model for any context. In that respect, research on the management of innovation has only just begun."

Joe Tidd *et al.* (2001)

Why do organizations look the way they do, and do they reflect the present most efficient structures and processes for a company's activities, or are they only relics from the past? Metaphors are often useful as images in our search for a better understanding of industrial organizations (Morgan, 1986). Using the body as a metaphor and image of an organization, the skeleton which keeps the whole body together can be compared to the formal organizational structures. The body movements that are managed by the bundles of muscles could then be an image of company work processes. But to keep the body moving we need the nervous system, which we could then compare to the informal organizational structures, like networks and other groupings — the company behind the chart. There is probably still an overemphasis today on the importance of a company's formal organizational structure, although its relationship to performance is still poorly understood. The above metaphor is to emphasize that all

131

these three elements must be considered in a holistic perspective when improved organizational solutions are sought.

The importance of process innovation to a company's ability to cut manufacturing costs, improve production efficiency, enhance product quality, etc., has been acknowledged for a long time (Skinner, 1992). However, little research has focused on how to actually organize these activities within companies, especially in the process industries. One should however be careful about copying well-functioning organizational solutions from other manufacturing industries into the R&D environment of the process industries. Referring to the above quotation, it is then not recommended to pick an "off-the-shelf" solution in an organizational development project, for although it can be seen as a quick fix, it will probably have no competitive impact and will not last very long.

Because product and process innovation in the process industries are characterized by an intertwining of development work, a project may cross over the organizational lines of marketing, R&D and production many times during its progress, as well as being defined as both product and process innovation. This chapter will start by reviewing various organizational concepts and follow up with by some empirical findings about the functional organization of process innovation. Finally, cross-functional teams and networks will be briefly discussed as complementary organizational solutions.

7.1 Corporate organizational structures — an introduction

R&D organizational structures are often similar to, and always embedded in, corporate organizational structures, although their inherent activities are not of a similar nature. It is however important, as illustrated in Figure 7.1, that the R&D structures fit the internal and external corporate environment and also respond to changes in those environments. The chances of a good R&D organizational fit instead of a misfit depend on how well structures, work processes and networks are dynamically integrated and harmonized within the corporate organization. Because of frequent changes (Bergfors and Larsson, 2009) in the internal and external corporate environment, the structural configurations of

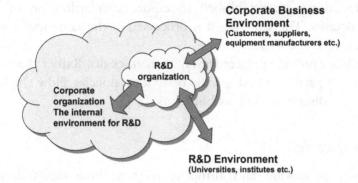

Figure 7.1 An R&D organization as an embedded part of a corporate organization.

today's R&D organizations are sometimes unfortunately fossils from the past and are as such often dysfunctional.

In discussions about the design of industrial organizations, the following structural dimensions and concepts are useful not only for organization theory researchers, but for industry professionals as well (Robbins, 1983).

Complexity

- Horizontal differentiation: The degree of differentiation between units based on the orientation of members, the nature of the tasks they perform, and their education and training.
- Vertical differentiation: The depth in the structure. Differentiation, and hence complexity, increases with an increasing number of hierarchical levels.
- Geographical (spatial) differentiation: The degree to which the location of an organization's offices, plants, and personnel are dispersed geographically.

Formalization

- Rules are explicit statements that tell an employee what he or she ought or ought not to do.

- Procedures are established to ensure standardization of work processes. The same input is processed in the same way, and the output is the same.
- Policies provide greater leeway than rules do. Rather than specifying a particular and specific behaviour, policies allow employees to use discretion, but within limits.

Centralization

- Decision-making and authority refer to how much decision-making is concentrated at a single point in the organization. A high concentration implies high centralization, whereas a low concentration implies low centralization or what may be called decentralization.

Starting with a clarification of the above concepts, one can design a number of different kinds of organization, which should preferably reflect the needs of the different corporate environmental situations symbolized in Figure 7.1. To exemplify just a few of them, we can take the organizational solutions presented by Mintzberg as "structure in fives" (Mintzberg, 1999). The presentation of the five structures will however rely heavily on Johnson & Scholes (1999):

- The simple structure: This is in many senses a non-structure. Few of the activities are formalized, and it makes minimum use of planning. It has a small management hierarchy, dominated by a chief executive (often the owner) and a loose division of work.
- The machine bureaucracy: Often found in mature organizations operating in markets where the rate of change is low. This configuration is still very appropriate for those producing commodity products or services where cost leadership is critical to the organization's competitive performance.
- The professional bureaucracy: This lacks the centralization found in the machine bureaucracy. Professional work is complex, but it can be standardized through ensuring that the professionals operating in the core have the same core knowledge.

- The divisionalized form: The important organization design issues are concerned with centre/division relationships. In particular, the corporate centre will specify levels of performance output expected from divisions or subsidiaries. In organizations selecting strategic control, the specifications of outputs are more likely to be expressed as a series of performance indicators, such as market share, efficiency ratios, etc.

- The adhocracy: This is found in organizations whose competitive strategy is largely concerned with innovation and change. The style of a corporate centre which fosters knowledge creation and innovation is of particular interest.

One can say that the professional bureaucracy is an organizational form that in many instances could characterize a traditional R&D organization in the process industries.

So whereas the machine bureaucracy relies on authority of a hierarchical nature — the power of office — the professional bureaucracy emphasizes authority of a professional nature — the power of expertize (Mintzberg, 1999). It is thus evident that the production organization in the process industries, which can often be characterized as a machine bureaucracy, does not necessarily operate without friction with an R&D organization that more resembles a professional bureaucracy or even an adhocracy, even if the latter organizational structures for R&D are the most popular.

The functional or departmental organization is still most common for production, sales and marketing, and R&D in many sectors of the process industries, but is sometimes complemented with work processes and networks. There are often many different kinds of activities that are carried out within the R&D department, of which many are not really R&D but are best located within this organizational framework. It is a rather interesting phenomenon in industrial organizations that tremendous energy is devoted to the construction of the best functional organization and hierarchy in the old tradition of scientific management (Taylor, 1916). In the day-to-day operation of this organization, however, much energy is subsequently directed to overcoming those functional barriers. One may even raise the question of whether today's functional organizations are approaching obsolescence.

To overcome the disadvantages of crossing these barriers, project organizations have since long been developed that cross the functional interfaces. Some companies can even be described as "project-controlled", because the work going on consists of just a bundle of very large projects. The matrix was later on created to overcome the disadvantages of all-too-rigid project structures, impoverishing the functional line organizations. A matrix organization is nowadays still quite a common solution that captures the best features of functional and project organizations. There are many different kinds of such matrix organizations that can be characterized by the location of the centre of gravity between the two organizational alternatives. In a study of alternative organizational solutions for product development, five such structures were assessed in a large survey (Larson and Gobeli, 1988): The functional, functional matrix, balanced matrix, project matrix and project team. The results from the study caution against the functional organization or the functional matrix or even the balanced matrix. Strong support was however evident for either the project matrix or the project team, of which the last will be further discussed in a following section.

Geographical differentiation — and decentralization

Two important aspects of functional organizations, especially in the perspective of globalization, are geographical differentiation and decentralization. There are unfortunately no such things as good or bad organizational solutions, but rather organizational solutions that have certain advantages and some disadvantages. The advantages of a geographically differentiated organization for various kinds of innovation activities in the process industries can be briefly summarized as follows:

• Good for application development; nearness to the customer.
• Good for product development; listening to the Voice of the Customer.
• Good for process development; close to production.
• Good for applied research; close to external knowledge centres.

The disadvantages of geographically differentiated organizations may on the other hand also be:

- Scattered resources.
- Geographical barriers.
- Less confidentiality.
- Difficult co-ordination.

Another important aspect and danger of decentralized organizations has been well put by Matheson (Matheson and Matheson, 1998):

> "What companies with a functional emphasis typically do is decentralise R&D by moving it closer to the business units. In the absence of strategic-decision analysis, it is likely that neither the business unit nor anyone else will be accountable for long-range renewal. Instead the overwhelming need to meet the numbers will result in quarter-to-quarter and year-to-year planning. At best, it will probably mean little more than a possibility of a tactical win — with a long-term failure."

In line with this reflection, Tidd & Pavitt conclude that decentralizing R&D reduces the scope for exploiting the interrelatedness of technologies, and that the most appropriate balance between corporate R&D by the divisions will depend on the strategy and technology (2000).

Centralization and decentralization of product and process innovation have been studied in the process industries in three case studies from the metal, forest and food industries respectively (Bergfors and Larsson, 2009). They illustrate the importance of moving away from sweeping generalizations of what R&D is and how it should be organized (different kinds of R&D). The results show that it is important to recognize that product innovation and process innovation may be organized differently with respect to centralization and decentralization. It is thus important to consider such dual structures of different kinds of R&D in organizational design, because the different solutions favour different kinds of innovation output. The conclusion is that a company wishing to generate radical product innovations should go for a more centralized organization. With regard to process innovation, a similar conclusion is drawn and is illustrated with the following statement by an R&D manager (Bergfors and Larsson, 2009):

> "as most of R&D employees are working very close to the production process we have to avoid ending up with yet another production engineer ... The autonomous R&D employees should contribute to process improvements which are more significant and long-term than the ones that the production engineer delivers".

Networks

The complex relations and communications in a large firm are illustrated in Figure 7.2 as a functional structure, but more as a relational, informal network. In the review of company organization and communication related to this figure there are a number of questions of interest to investigate: Is there a relation? What type of relation? How strong is the relation? Is there a need to change and improve this pattern?

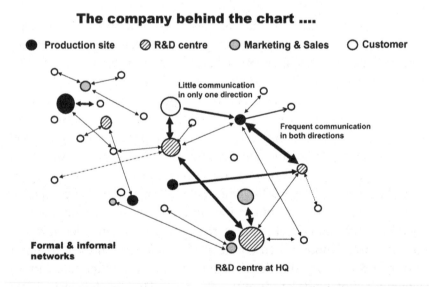

The company behind the chart ....

● Production site　⊘ R&D centre　◐ Marketing & Sales　○ Customer

Little communication in only one direction

Frequent communication in both directions

Formal & informal networks

R&D centre at HQ

Figure 7.2 The company behind the chart as formal and informal networks. The communication between different organizational units is symbolized with arrows of different thickness.

As the number of company R&D sites at home and abroad grows, R&D managers will increasingly face the challenging task of co-ordinating such structures and related networks. More than being managers of people and processes, they must be managers of knowledge (Kuemmerle, 1997). As an example, Kuemmerle discusses different objectives for geographical differentiation. He points out that R&D managers will increasingly face the challenging task of being co-ordinators, not local administrators. In his large study the decentralized R&D sites are differentiated into two categories:

- Home-base-augmenting site, established to tap knowledge from competitors and universities around the globe; information flows from the foreign laboratory to the central laboratory at home.
- Home-base-exploiting site, established to support manufacturing facilities in foreign countries or to adapt standard products to the demands there; information flows to the foreign laboratory from the central laboratory at home.

Virtual organizations (which however will not be discussed further in this book) are characterized by high flexibility, quick responses, fewer interface problems and better knowledge integration (Bowonder *et al.*, 1995). The ability to expand core skills is one element in virtual organizations where the effective apparent boundaries of a firm are expandable when and as required. For a thorough review of R&D strategy and organization and a discussion of the balance between centralization and decentralization, geographical differentiation, organizational form of technological collaborations and technology acquisition, see Chiesa (2001).

Organizational design

Finding well-functioning organizational solutions for R&D is not an easy task, but in the end often rewarding for the people concerned. Since each company and company organization is unique, organizational improvements and changes should be unique (Woodward, 1965). Referring to the introductory quotation from Tidd, the practice of organizational design is still however on a low level, lacking knowledge and a fact-based platform. The consequence is well put by Nadler and Tushman, as: "for every design that displays a spark of genius or a truly new insight into social organisation, there are ten misguided efforts that reflect no more thought than a couple of sketches on napkins over lunch" (Nadler and Tushman, 1997).

A better approach to organizational design is to start by acquiring a deeper understanding of the many operational aspects of an R&D organization, and base new solutions on the collection and analysis of factual information in this context. The ambition should further be to find solutions that fit the company's internal and external environment in consensus with the people affected by such solutions. Figure 7.3 may serve as a guide on how to proceed in the design of an organization. The input, i.e. the determinants for selection, may of course be of different kinds, but considering aspects from the following areas could be a fruitful starting-point (Hall, 2002, p. 82).

- Company strategy as the determinant?
- Company size as the determinant?
- Company technology as the determinant?
- Institutional explanations?
- Expected output?

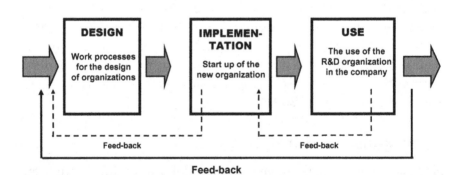

Figure 7.3 A simplified model for designing an organization. The input for the design is the determinants previously presented as well as the expected outcomes. The output is measurables of organizational performance. The intermediate phase, implementation, is inserted to emphasize this important but often neglected part of organizational change.

Since each organizational solution has its own advantages and disadvantages, we come back to the question of what is the purpose of an organization, or rather the purposes of organizations.

The following different and alternative outputs may be expected from an organization, and can then also be looked upon as the expected output (objectives) for a selected organizational design and as such hopefully enablers for better performance. Such outcomes can be classified (Bergfors, 2009) as strategic (focusing on effectiveness) or organizational (focusing on efficiency):

- Good control (power?)
- Efficient decision processes (good management)
- Better corporate information flow (good communication)
- A good framework for effective work processes.
- Easy internal collaboration.
- Easy external collaboration.
- Improved innovativeness.
- Improved personal and organizational learning.
- A good context for the corporate culture.

Depending on what one would like to improve in an R&D organization, solutions that can be expected to deliver those outcomes are sought. Since the purpose of the R&D organization is to facilitate work, not to hinder it, it is vitally important that selected solutions and changes are well accepted by the people who work in the organizations. Because of that it is advantageous if the R&D organization actively participates in the process of change.

An evolution of the corporate organization is sometimes preferable, and the same goes for changes in the R&D organization. Drastic changes in the corporate R&D organization, which we usually call re-engineering as opposed to continual improvement, may however sometimes be necessary (Kotter, 2007). Depending on if the necessary organizational change is of an incremental or radical character, the necessary time frame may have to be gradual or rapid. Changes may however not be of a linear character in time (punctuated change) which can make transformation of organizational structures complex (Sastry, 1997).

7.2   The functional affiliation of process innovation — evidence from the European research project

While most companies still adhere to some form of functional structure based on common purpose or contribution to the larger organization (Nohria, 1995), it is not uncommon for mixed forms to be used. There are basically three ways that the innovation of process technology can be arranged in firms in the process industries which have a functional organization:

- Process innovation can be organized and managed within the R&D department together with product innovation activities.
- Process innovation can be organized and managed within the production organization.
- Process innovation can be organized within both the R&D department and the production organization.

An organizational framework

When we examined the process industries in Chap. 4, we found that the distribution of resources for process development leaned more towards development that can be classified as improvements of existing production processes rather than development geared towards breakthrough process technology (Lager, 2002). This is however a strategic choice that the firm has to make. In general, research findings suggest that radical innovations require the technical knowledge and slack resources normally available to larger and more complex organizations (see review by Damanpour, 1996). This is because organizations pursuing radical innovations need to be able to summon up the necessary human and technical resources and absorb the higher cost of failure of such innovations.

Other studies of radical innovation have found that it was more likely to emerge in centralized structures (Ettlie *et al.*, 1984) and when separated from ongoing business activities (Rice *et al.*, 1998). Considering organizational affiliation specifically, Lawrence and

Lorsch (1967) proposed that the more organic organizations (meaning decentralized and less formalized), which are often seen as characteristics of R&D departments, are better at achieving uncertain tasks, such as radical innovations.

More mechanistic organizations (meaning hierarchical and more formalized), often more typical of the production organization, are on the other hand better geared to certain tasks, such as incremental innovation (Lawrence and Lorsch, 1967).

This notion can also be linked with the fact that the R&D department and the production organization have different goals and reward structures that support different types of activities. Utterback for example argues that since the R&D department is geared to focus on long-term goals, it has a propensity for turning out more radical innovations, while the production organization focuses more often on incremental improvements for measurable benefits in the short term (Utterback, 1994).

Nevertheless, a more recent study of innovation in the pharmaceutical industry found that incremental and radical innovation should not necessarily be managed differently (Cardinal, 2001). While there have been no studies specifically considering the organizational affiliation of process innovations, there are several arguments as to why affiliation may be related to the degree of newness of process innovation.

Pisano (1997, p. 155) noted that there were no previous studies linking innovation performance with organizational structures specifically for process innovation. In his study of pharmaceutical and biotechnology companies he identified two models for organizing process and product R&D; Pisano labelled these as the "integrated model", which describes process innovation organized in R&D, and the "specialised model", which describes process innovation organized in both R&D and production, with a handover in between. However, as product innovation in his book is defined as a desire to improve finished products, the process innovation activities are actually more akin to product innovation, as they deal with the development of improved product properties. Thus in the specialized model, process innovation can actually be said to be organized in production. In studying the integrated and specialized models, Pisano found several differences in process innovation lead times and costs. While these results were not statistically significant, they suggest that pros and cons exist for organizing process innovation either in the R&D department or in production.

Evidence from the European research project

The following results are from the previously presented European research project (see also App. C). The results are presented by Bergfors (2007), and will also appear in a forthcoming paper (Bergfors and Lager, 2011).

Newness versus organizational affiliation of process innovation

The future organizational affiliation of process innovation cross-tabulated against different degrees of newness of the innovation is shown in Table 7.1. The null hypothesis that the two variables are independent was tested with a Pearson chi-squared test.

The result indicates that the null hypothesis can be rejected, and hence that the two variables are not independent (p-value = 0.008). The structure of the association can be summarized as follows: in the "Very High" and "High" categories of newness of process innovation we find an over-representation of companies organized in R&D compared to the expected

Table 7.1   Future Organizational Affiliation of Process Innovation for Different Degrees of Newness of Process Development.

Newness of process innovation	Future organizational affiliation of process development			
	R&D (%)	Production (%)	Both (%)	Total (%)
Very high (<32.5%)	66.7	19.0	14.3	100.0
High (32.5–50.0%)	63.0	29.6	7.4	100.0
Low (50.0–70.0%)	52.4	38.1	9.5	100.0
Very low (>70.0%)	6.3	68.8	25.0	100.0
Total	50.6	36.5	12.9	100.0

number. In the "Very Low" category we find an over-representation of companies organized in production compared to the expected number.

In view of the empirical findings we can suggest that if a company pursues process innovations of high or very high newness, which are labelled *radical* innovations for the rest of the discussion, the recommendation is that process innovation be organized within the R&D department (Bergfors and Lager, 2011). On the other hand, if a firm pursues process innovation with a very low level of newness labelled as *incremental* innovations, these activities ought to be undertaken within the production organization. There are also a number of companies which pursue both radical and incremental process innovations and which have the option of organizing these activities in both R&D and production. Although the results are clear on how to organize process innovation, we still have no empirical evidence to suggest why. This is however further discussed in the following sections.

Pros and cons of undertaking radical process innovation in the R&D organization

The empirical research findings show that companies who wish to pursue *radical* process innovations would do better to organize this activity within the R&D department. Some different rationales for organizing and managing radical process innovation here could be:

Advantages

- Economies of scale in research: These can be reached by pooling technical expertize, scientific knowledge and other resources within R&D. Ettlie *et al.* (1984) found that a growing number of technical specialists seems to promote the adoption of advanced technology. Pooling staff also simplifies knowledge-sharing among research professionals and makes it easier to attract new talented employees to process innovation.
- Distance from everyday operations: Being organized within R&D means that R&D employees can detach themselves from the everyday operations and problems in the production plant.

The latter often require immediate attention and can divert attention from non-time-critical research. Another aspect of keeping distance is that radical process innovations often involve big changes in how work ought to be done in the production organization. If process innovation is organized under production, there might be more barriers to introducing innovations that change the immediate working environment. Studying metal-making, Moors and Vergragt (2002) found that a high degree of technical and organizational intertwinement often complicates the implementation of more radical innovations.

- Strategic control: This often becomes easier if process innovation activities take place under R&D. For example, Chiesa (2001) suggests that the more centralized a function, the easier it is to control for strategic direction. If process innovation is organized under production, it is more likely that production issues will have higher priority than strategic innovation goals.
- Longer time frames: R&D departments usually have a longer time horizon for investments and projects than production (Lawrence and Lorsch, 1967, p. 36). Production is more focused on completing projects with short-term payback.
- Integration with product development: Conducting process innovation as part of R&D makes closer integration with product innovation and other R&D activities more straightforward. This is especially important when product specifications cannot just be "thrown over the wall" for further development of process technology, or when product development and production process development processes are closely intertwined (see especially Lim *et al.*, 2006 and Pisano, 1997, for examples from the pharmaceutical industry).
- Risk acceptance: Production units are less inclined to make large investments with uncertain payback. However, radical innovations are by nature uncertain and are better nurtured where short-term goals do not compromise long-term undertakings. R&D often has different budget constraints to production, which encourages risk-taking to a higher extent.

- External relations: These are important for bringing new knowledge into the company (Cardinal, 2001). Studies in steelmaking suggest that heterogeneous and externally oriented relations facilitate the development of radical innovations (Moors and Vergragt, 2002, p. 295). Managing external contacts is often easier to do from the R&D department.
- Organization structure: R&D is often seen as less hierarchical, less formalized, and with a culture that permits free thinking. Organizations of this type are often referred to as "organic" and are geared towards uncertain tasks (Lawrence and Lorsch, 1967, p. 31), such as pursuing radical innovations.

Naturally there are also some drawbacks to organizing radical process innovation under R&D:

Disadvantages

- Lack of production understanding: When the R&D professionals have less knowledge of the production processes, they may have more difficulty in creating new radical ideas. In short, they do not know what the customer needs in terms of process innovations, and in these cases R&D can often be regarded as an ivory tower (Chiesa, 2001).
- Distance from actual problems: R&D personnel too far removed from the actual production processes do not see problems as they arise.
- Technology transfer: When a new process is developed, the technology transfer and project handovers are likely to be more difficult if it has been developed outside of the production organization that has to implement it. There are organizational, as well as purely technical issues to overcome. For example, changes in technology may call for new skills and changes in systems and procedures for production control and scheduling. New organization structures may be required. Furthermore, the "not invented here" syndrome may hinder process innovations originating from R&D from being adopted in production.

- Learning: Many organizations emphasize learning by doing through production experience, and that much important learning takes place on the shop floor (Hayes *et al.*, 1996, p. 104). If process innovations are developed away from the production organization, the chances for learning from plant operators are limited. This often causes problems during startup.

Pros and cons of organizing incremental process innovation in the production organization

The empirical research findings show that process industry firms who pursue incremental process innovations would do best to arrange these activities under the production organization (Bergfors and Lager, 2011). There are several arguments for doing this:

Advantages

- Closeness to customers: The production organization is the main customer for process innovations within the company. Being organized in production close to the customers means that process innovations can pick up on emerging needs and unique production characteristics (Cardinal, 2001). This encourages the alignment of process innovation projects with production needs.
- Knowledge of the specific work environment: Knowledge of the work environment and the production processes will facilitate process innovation. The firm must continually use the plant as a source of knowledge, or R&D employees may lose the specific knowledge of that environment (Pisano, 1997, p. 279).
- Increased accountability for budgets: As production is continually measured by key performance indicators, there is little room for big spending. As a result, innovation work within production is typically more fiscally responsible. Additionally, if an organization is being measured on costs, it will focus innovation work on cutting these, which is the prime objective of process innovation.
- Learning: If process innovation is conducted in the production environment, the operators can start their learning process

before the new process technology is implemented. However, this works best with technology that offers incremental improvements.

- Faster process innovation cycles: As the return on the investments needs to be high, there will be a tendency for production to focus on projects that will be quickly "up and running" for immediate profits.

- Improved technology transfer: Handing over technology and implementing innovations in the production process is simplified when development work is carried out at the production unit. Problems can be solved on-site by people who understand the context.

- Straightforward communications: Studies show that while radical project success is facilitated by cross-functional teams, incremental projects are easily burdened by information overload created by cross-functionality (Cardinal, 2001).

- Organizational structure: Production is often seen as a mechanistic organization, where work is more closely supervised and formalized, and whose main objective is to solve specific tasks (Lawrence and Lorsch, 1967) — a fitting environment for incremental process innovation.

There are also some drawbacks to incremental process innovation in the production organization:

Disadvantages

- Duplication of research: Research efforts may not so easily be duplicated across production plants within a large corporation when each production unit only looks out for itself. When process innovation is conducted in production, the close focus on specific problems may mean that vital ideas and technologies are not communicated to other parts of the organization.

- Focus on "safe" projects: Being measured on success of projects and cutting costs in production will drive innovators to choose simple and safe projects instead of more difficult but also potentially more rewarding projects.

- Loss of strategic direction: When development projects are chosen on a project-by-project basis based on day-to-day production
  needs, there is a risk of long-term objectives being overshadowed.
  Hayes *et al.* argue that process technology choices should not be
  solely the domain of technical and financial specialists: "Even if
  each individual investment decision seems sound from a technical
  and financial standpoint, they may still not fit with the company's
  operating strategy" (Hayes *et al.*, 1996, p. 99)

Organizing process innovation under R&D and production simultaneously

A substantial percentage (12.9 percent) of the R&D managers undertaking the survey predicted that process innovation would in future take
place in both the R&D department and the production organization.
Separating different types of innovative activities in different organizational units could imply organizing radical process innovations under
R&D, while incremental process innovation occurs under production.
This alternative could offer the company an opportunity to have the
best of both worlds, as pointed out by Clark and Fujimoto (1991,
p. 123): "Effectiveness of process engineering depends as much on the
ability to interact with product designers and the factory as it does on
technical skills". (The reference was intended for process engineering,
which is somewhat different from process innovation, but can still hold.)

However, there are also several difficulties concerning integration
and communication issues between R&D and production that are not
easily overcome. For example, a high degree of integration is necessary to inform R&D employees about technical problems, and to
acquaint production personnel with new processing techniques that
will be introduced (Lawrence and Lorsch, 1967, p. 45). Also, R&D
and production must agree on what technological innovations are
possible and desirable. The overlap must be managed carefully, as the
design of products has a major effect on the cost of making them and
the time-to-market (Pisano and Wheelwright, 1995). However, split-
ting up process innovation, which is often a small organizational unit,
may cause a dangerous loss of "critical mass" as well as creating a new
communication problem.

Some companies employ cross-functional teams in process innovation as well as product innovation. In a sense, the decision on organizing process innovations can be taken on a project-by-project basis. Some projects may be more suited to be run under production, while others need the resources and special expertize found in the R&D department. Based on the degree of newness (which can range from incremental to more radical), and referring to the survey and literature, the advantages of a case-by-case approach locating process innovation projects to the R&D organization might be to give a good environment for more long-term process innovation projects of a more radical character. The advantage of assigning process innovation projects to production might be to facilitate incremental development by making it easier to transfer results directly to production.

All this implies that the departmental affiliations may be looked upon as less important than the organization of efficient work processes for process innovation transcending departmental demarcation lines; this will be discussed in detail in Chapters 8 and 9. The shift from a strict departmental organization to multifunctional teams headed by project managers has been a clear trend for product innovation in other manufacturing industries (Clark and Fujimoto, 1991). Increased utilization of cross-functional teams has been found in the process industries (Chronéer, 2003). Nevertheless, the importance of establishing multifunctional structures and modes of collaboration for process innovation and production is still relevant, because the departmental organization will probably survive for some time in the process industries.

Some sort of organizational backbone will be needed even in the future, although the focus may be more on work processes, including in this case the process innovation work process. However, the ownership of such a work process will continue to be held by a selected functional unit also in the future.

7.3 *Ad hoc* cross-functional teams

The introductory quotation from Hans Rausing at the beginning of this book makes it clear that research and technical development are risky by nature, sometimes carried out as more or less non-repeatable processes. The people at R&D do certainly have a difficult task, always aiming at the future and fighting for resources. Other company employees, and certainly company management, must respect the unique nature of R&D and give its practitioners trust and confidence in their difficult work — the "laboratory life" (Latour and Woolgar, 1986). For further reading on the important subject of managing people in an R&D and engineering environment, please see Cleland and Kerzner (1986) and Sapienza (1995).

Complementary organizational solutions to the structural organization and work processes have been used for a long time in industry. Some of them are however probably more a "confession of the lips" than solid, well-implemented organizational solutions. The recording of the following mission statement within a business unit by a director of R&D in the food industry will underline this state of affairs: "I have a dream: development by research, engineering and marketing" (Moenaert *et al.*, 1994). It was then also noticed that R&D personnel could assume marketing responsibilities like contacting and informing customers and that there also are some industries where marketing personnel with a strong technical background could assume R&D responsibilities.

Cross-function organizational solutions have thus been advocated and used for a considerable time, but mainly in product development. Such corporate behaviour is however easy to wish for but often harder to establish. An early publication, from the vast literature in this area, gives the following quotation from the John Deere corporation (Lorenz, 1991):

> "We'd moved beyond the stage that many companies have still only reached today, of drawing new lines between boxes but still having a matrix. A lot of so-called simultaneous or concurrent engineering groups are nothing more than official format. Specialists of every type were transferred to what Wyffels calls cradle-to-grave project teams, working together from initial concept until well after product launch."

Referring to the last sentence in the quotation, this is a state of affairs, which hopefully now many firms have reached for product development, is still probably worth reflecting upon for other innovation activities like process innovation. The need for learning through relationships, and for engineers to have a broad systematic competence profile that facilitates cross-functional communication within and across corporate borders, has been stressed by Harryson (1997). Cross-functional groupings can also be called interdepartmental integration, and different modes of such integration have been well described by Kahn (1996):

"The interaction philosophy favours communications between departments, which encourages managers to hold more meetings and establish greater information flows between departments... elaborate meeting schedules and extensive information networks for the routing of standard documentation.

In the collaboration philosophy, continuous relationships between departments are stressed, not just transactions between departments. There is an emphasis on the strategic alignment of departments through a shared vision, collective goals and joint rewards, along with emphasis on informal structure, to manage relationships."

The results from the survey carried out by Kahn indicated that a collaborative mode had a strong positive influence on product development performance.

Four networking mechanisms were recommended by Harryson (1997); in a slightly modified version they could also be an excellent recommendation for process innovation in the process industries:

- Strategic training and rotation of engineers to ensure that every member of the innovation process has holistic understanding of the needs of technology, the needs of the market, and the requirements of manufacturing.
- Focus on prototyping in the research stage encourages researchers to exchange knowledge with manufacturing up-front (market-driven research).
- Direct transfer of flexible researchers to the factory floor drives technological competence into production processes and products.

- Extensive networking with external centres of excellence and key suppliers replaces excessive internal research efforts by overspecialized researchers who sometimes refuse to leave their labs.

One important objective for future organizational design is that the selected solution should stimulate organizational learning, which has been well presented and defined by Hörte (1997):

> "Organisational learning can be viewed as changes of the common mental models within an organisation. A common mental model defines the way members view their organisation (What are we here for? What are we doing? How do we evaluate what we are doing? etc.) and how it is related to and viewed by the surrounding world. (What do they say about us?).

> A common mental model is therefore a global perception in which one's own organisation is part of a larger global picture. Common mental models are manifested by the way organisations handle internal and external relations, the organisational routines, standards and values. Organisational learning processes can be defined as changes in the common mental models within an organisation."

In Chapters 8 and 9, work processes for process innovation are presented in depth, and one of the most important aspects of these is to stimulate and improve organizational learning. This in turn becomes an important objective with more efficient work processes in order to deliver lean process innovation as an outcome.

7.4 Summing-up and some issues to reflect upon

In discussions of strategy and structure (Chandler, 1962) it is sometimes debated which should come first. In this presentation it is felt natural that strategy is the starting point for how forces should be deployed (von Clausewitz, 1984). This chapter raises some general issues, arguing that wise organizational solutions are effective tools because they are difficult for competitors to imitate. The formal organizational structures are further discussed, and it is pointed out that solutions from other manufacturing industry or even other companies in the same sector may not be applicable to a company's specific needs in the process industries.

The research results presented in this chapter provide empirical evidence that there is a positive association between the degree of newness of process innovation pursued and the organization of innovation. More specifically, this study shows that if the company's newness of process innovation is high, an organizational affiliation to R&D should seriously be considered. If, on the other hand, the newness of process innovation within the firm is low, affiliation to the production organization is an alternative that is recommended.

However, the findings should not be considered as strictly normative. Some other possible rationales for the organizational design of process innovation could be taken into account, and these are presented and discussed in depth. Increased economies of scale, distance from everyday operations, increased strategic control, and higher risk acceptance are some rationales why process innovations should be organized within R&D. Meanwhile, the knowledge of the specific working environment, increased accountability for production budgets, faster learning and improved technology transfer are some reasons why incremental process innovation will benefit from being organized as part of the production organization.

There are no such things as off-the-shelf organizational solutions of excellence, and the recommendation is to base new organizational design on more factual information and operational needs. It is proposed that the special demands on the R&D organization very likely call for different solutions than for a production organization because of the different character of the inherent activities. The globalization that is now affecting most sectors of the process industries is briefly touched upon from the perspective of differentiation and decentralization, but needs careful consideration in the company's organizational design process.

- *Using Mintzberg's "structure in fives", how would you like to characterize your firm's organization and R&D organization?*

- *In the discussions of alternative organizational solutions, has your firm considered more fundamental questions such as what kind of output should expected be from different organizational solutions?*
- *How much consideration is given to the organizational implementation phase when organizational changes are made in your firm (see Fig. 7.2)?*
- *How well is your R&D organization integrated in the overall corporate organizational structure?*
- *If your company is operating over many national boarders, how well are your firm's innovation activities adapted to such global operations?*
- *Have complementary organizational solutions for R&D been considered systematically, or is the company stuck with the traditional one-and-only functional organizational chart?*
- *Have you considered the proper functional affiliation of process innovation and the pros and cons of alternative solutions?*
- *Starting with the functional organization of your firm's process innovation, benchmark the list of pros and cons and discuss possible improvements in your R&D organization.*
- *Are you creating cross-functional teams in your firm's process innovation projects?*

Chapter 8

A Conceptual Model for the Development of Process Technology in the Process Industries

"If you can't describe what you are doing as a process, you don't know what you're doing."

W Edwards Deming

One way to improve the performance of process innovation might be through a better understanding of the work processes. In other manufacturing industry, improved performance in product development has often been achieved with new work processes commonly referred to as concurrent engineering, simultaneous engineering or integrated product development. These work processes integrate the whole organization at an early stage in the development process, using multifunctional teams with a strong customer focus.

In the development of new products at a faster pace and at lower cost, the formalization of the product development process has often gone far. When the process industries start to formalize work processes for product and process development, it is important to learn from industrial experience and academic research about the product development process in other manufacturing industry, but also to identify the specific needs of the development environment for the process industries. The importance of developing well-functioning work processes for different innovation activities has been expressed by Heygate (1996):

> "The different treatment accorded product and process innovation by many companies stems from a belief that the two have fundamentally different strategic value. Process innovation is clearly the poor relation. But in some industries, entrepreneurial managers are learning that product and process innovation are of equal strategic importance, requiring the same approach to research, development and continuous improvement."

Publications in the area of process innovation work processes are however still scarce, and the following text is mainly based on one of the first publications (Lager, 2000), in which the structure of such a work process was outlined. Further on, empirical evidence of the use of such a work process in the process industries is presented and integration with other work processes is discussed.

In a later publication (Lim *et al.*, 2006), and in reviewing such a work process and its potential use for process innovation in biotechnology, the conclusion is that simultaneous development of new products and production process technology is an issue of growing importance for this kind of industry. But the importance of the timing of each kind of development activity is also emphasized, as well as the difficulty of striking a proper balance and good integration.

8.1 Mapping corporate innovation work processes

In the metaphor presented in the introduction to this part of the book, company work processes were compared with the bundle of muscles that give the body motion. Current company working practices, nowadays often called company work processes, do not however seem to resemble well-trained and agile muscles, and are seldom designed to satisfy the company's needs of today or tomorrow; they are often only regrettably an organizational memory of dubious quality from the past.

There are many descriptions and definitions of the work process concept. An old but still valid and useful definition from AT&T (1988) states:

"A process is the set of interrelated work activities that are characterised by a set of specific inputs and value-added tasks that produce a set of specific outputs... Processes may consist of a collection of sub-processes."

Many activities in industry can be looked upon as work processes. The work process must have both a customer and a supplier, and the existence of feedback loops is illustrated in Figure 8.1.

Melan (1989) characterizes four different types of transformations occurring in different types of work processes as: physical, locational, transactional and informational. Other characteristics of a formal work process are that an "owner"of the process should be recognized and that the process should have clear boundaries (interfaces) to connecting work processes. The term "development process" is however deceptive, because it gives the impression of a completely ordered linear and logical process that is controlled step by step. Those who have been directly involved in development activities can

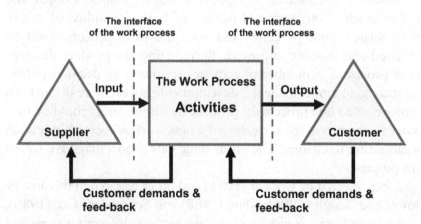

Figure 8.1   A symbolic model of the work process (after AT&T, 1988).

however testify that the development process is often of a very complex or sometimes even chaotic nature.

Utterback (1974) presented a simple three-phase model for the development process:

- Generation of an idea: Involves synthesis of diverse (usually existing, as opposed to original) information, including information about market and technology.
- Problem-solving: Includes specific technical goals and designing alternative solutions to meet them.
- Implementation: Manufacturing, engineering, tooling, plant startup, and market launch required to bring an original solution or invention to its first use or market introduction.

Booz, Allen and Hamilton (1982) presented a simple linear model of product development, including the steps: idea, preliminary assessment, concept, development, testing, trial and launch. At that time many companies may not have formalized their product development efforts, although that should not *a priori* be confused with the fact that many companies had established routines and knowledge about how to develop new products.

Research in this area by Cooper (Cooper, 1988b; Cooper and Kleinschmidt, 1986), gives a picture of both the industrial use of more formal work processes and how the work process could be designed and possibly improved. Recognizing the product development process as a number of activities separated by decision points, the stage gate model made the development process more distinct. To improve an ad hoc process for product development, it could be necessary to overstress the formality of a new work process to induce an organization to change. The linear stage gate model may have served this purpose.

Later research by Cooper (1994) and other research in this area by Bower and Keogh (1996); Kline (1985) and Schroeder *et al.* (1986), gives a more complex picture of the product development process and indicates that future work processes must be more flexible and adapted to the individual characteristics of different development projects.

This research also tends to take a more sceptical view of the product development process as a well-controlled linear process, and descriptive research and case studies sometimes present a picture of a very chaotic development process. This observed fact, however, does not necessarily preclude the possibility that this chaotic process could be made less chaotic and that the output from development processes could be better if they had had a more formal structure.

Pisano (1997) recognizes the importance and competitive position of the development of process technology. Several approaches for process development are introduced, recognizing the plant as the customer to the process and the importance of letting process development go hand in hand with product development in product development projects. How much structure should the innovation process have, and how formalized should it be? The answer proposed by Norling (1997) is that it depends on the type of work that is going on and the type of organization the activity is being performed in. Re-engineering of R&D work processes can probably be a powerful tool for R&D in the future, according to Brockhoff *et al.* (1997); Hammer (1990) and Liebeskind (1998).

For further reading about work processes see, for example, Hammer (2007); Malone *et al.* (2003), Margherita *et al.* (2007) and Grönlund *et al.* (2010). Even then, restructuring and improvement of R&D work processes must start with a better definition and description of the present development processes.

8.2 A conceptual model of the process innovation work process — evidence from the European research project

The following results are from the previously introduced European research project, which appears in App. C. The following text and model were initially based on the author's own practical experience with the development of process technology and on fragments from a number of publications and discussions on the subject of development and innovation. The text and the model were then discussed with R&D managers from the process industries. In those discussions it was understood that the concept of a process development process was still much associated with this chain of practical test work in laboratories and in pilot plants, focusing on the test environment, which will be presented in the following section.

A traditional model of the process development work process

The discussions resulted in some modifications and refinements of this model, which was designated the traditional process development work process and which is a kind of descriptive picture of how practical test work is done in the process industries today. In Figure 8.2 the four steps in process development have been put together, giving some sort of general simplified model of the traditional process development work process, actually focusing more on the different environments for test work but not giving a very complete picture of a total process innovation work process.

Process development sometimes includes only one or two of these steps, sometimes in a different order and also with different types of iterative loops. Each development step serves different purposes in the total Process Development Work Process. Short cuts can be taken;

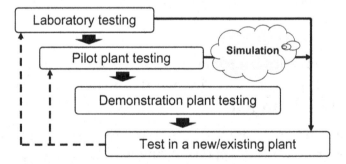

Figure 8.2   Different kinds of test environments in the process industries. Process innovation does not necessarily involve all the steps shown, nor are they necessarily performed in the order shown. Several iterative loops are common. Simulation can be a supplementary activity or sometimes an activity that replaces practical test work. After Lager (2000).

laboratory results can for example be scaled up and applied directly in the production plant; simulation can replace pilot plant testing. The model includes the following four parts:

Laboratory testing

Laboratory testing is often considered as the starting point for process development and includes experimental work in the laboratory in batch testing or in more continuous laboratory testing. Most of the experimental work is often done in a laboratory environment because of ease of experimentation and low costs. The laboratory testing part may have different objectives depending on the character of the development work. Development methodologies can certainly be improved in laboratory testing; one important step is the introduction of methodologies for experimental design and multivariate experimental evaluation, as well as other optimization and simulation methodologies.

Simulation

Simulation has already replaced test work in some applications, and will in many respects revolutionize process innovation in the future. Simulation can be used in all the steps presented in Figure 8.2 as a supplementary activity to testing in laboratories, pilot plants and operating plants. Well developed simulation programmes also have a very strong potential to be used in training people before and after the start up of new process technologies and new production plants. The performance of simulation and simulation programmes, however, depends to a large extent on a fundamental understanding of the inherent process technologies. The advancement of simulation technology today is at such a stage that it is not advisable to start any major new installation of process technology without a previous or simultaneous installation of proper equipment and programmes for simulation (Contreras *et al.*, 2005; Harding and Popplewell, 2000; Laganier, 1996; Shelden and Dunn, 2001; Young *et al.*, 2001).

Pilot plant testing

Going from the laboratory stage to the pilot plant stage is an important step to improving understanding of the process. The large increase of cost associated with pilot plant testing makes the balance between pilot plant testing and laboratory testing an important issue. One reason for moving the development process into the pilot plant is to test the full process as a closed system and to determine the influence of circulating loads on the process results. Another reason may be production of test samples in larger quantities for further testing. For an in-depth presentation of pilot planting, design and construction see further (Palluzi, 1992; Palluzi, 1997).

Trials in a demonstration plant

Trials in a demonstration plant are often very expensive and are carried out in a major development project when the process is of a completely innovative character and existing pilot plant facilities cannot be used. The primary objective of using a demonstration plant is not as an alternative to the pilot plant, but to run trials that complement previous pilot plant testing. These objectives may be:

- To test wear on the process equipment.
- To develop process control strategy and systems.
- To scale up the process to something between the pilot plant and the production plant.
- To produce larger samples of products for subsequent testing in the customer's process.

Production plant tests

Nowadays many tests are conducted directly in the production environment, giving fast feedback and reliable test results. Finding a correct balance between the goals of fast process development in plant tests and minimum production disturbance will be an important issue in the future. Production plant trials may be undertaken to test new equipment, new process conditions, new process configurations, etc. The role of the production plant as a source of innovation is of great importance, and it may also serve some other purposes in future integrated development work involving the production plant operational crew and the development teams. The use of the production plant for more development activities and improvement work may have future organizational implications and lead to new development models. There are also two fundamentally different cases:

- The process is to be implemented in an existing production environment or existing production plant.
- The process will be implemented in a completely new plant that must be built before production starts, a fact that made part of the unique organizational solutions possible at Chaparral Steel (Leonard-Barton, 1992).

Different situations also exist according to whether the new process is to be integrated into an existing production structure or operated as a stand-alone process. The traditional process development work process presented in Figure 8.2, however, lacks some of the fundamental parts that are often much more emphasized and articulated in descriptions of the product development work process.

Compared to the simplified general model of the development process presented by Utterback (1974), the first and second parts are missing since the ideas of generation phase and implementation phase are not explicitly stated. According to the author's practical experience, this is regrettably also often the case in the industrial development of process technology. In the following text a new conceptual model for the process development process has thus been developed that puts more focus on those missing parts, treating process development not only as practical test work.

A new conceptual model for the process development work process

Two assumptions for the development of the new conceptual model for the process development process are given below:

- *The process development work process is assumed to be unique for each company and possibly also for different types of projects. The consequence of this is that the development of a company-specific process development work process needs a simplified structural model that can be further developed into a more detailed company-specific model.*
- *The previously presented traditional model of process development lacks two important structural phases included in the model presented by Utterback.*

The simplified model of the general innovation process presented by Utterback (1974), including the three phases of idea generation, problem-solving and implementation, was considered a good structural starting-point for a new conceptual model of the process development work process. This model can then be developed in detail and customized by each company in the further development of its specific work processes. The building blocks of those detailed work processes for process development are probably rather different from the building blocks of the product development work process. The latter starts in the first phase with the identification of customer needs and ends with product commercialization or product launch. In the process development work process we must instead identify production needs in the first phase and end up with transfer of development results to the production process. This gives the conceptual model presented in Figure 8.3.

Phase 1 — Identifying internal production needs or product development needs

The idea of generation phase focuses on the identification of internal production needs. An analogy to the expression "listening to the

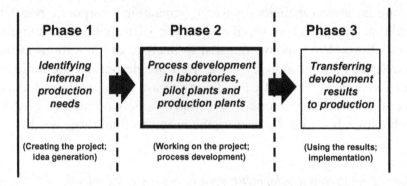

Figure 8.3 A conceptual model of the process development work process. Each phase comprises a number of sub-processes. The shaded Phase 2 in the middle represents all activities shown in Fig. 8.2.

Voice of the External Customer" in product development is "listening to the Voice of Production". This Voice of Production can possibly be developed into the same structural clarity as the Voice of the Customer (VOC) in product development, using proven Quality Function Deployment (QFD) methodology (Tottie and Lager, 1995); see further the presentation of QFD in Chapter 5. The reason that this phase is frequently neglected is probably because it is often confused with the experimental sub-process of doing laboratory testing. A way to improve this fuzzy front end phase in process development could be more desk research, brainstorming with production staff and workers, discussions with product developers, discussions with equipment manufacturers and suppliers of chemicals and reagents, structural analysis of the production process, and exploratory laboratory tests.

Sources of technical innovation for a chemical industry are discussed by Hutcheson and colleagues, who emphasize how important it is for R&D in the process industries to collaborate with external industry like process contractors, equipment manufacturers and raw material suppliers (Hutcheson *et al.*, 1995; Hutcheson *et al.*, 1996). Freeman (1990) lists 15 possible sources of new technology, of which in-house R&D is only one. He further reflects, however, that "the most successful and innovative firms have been characterized by a strong in-house capability for R&D, generating a corporate research tradition and company-specific processes of technology accumulation". Enos (1962) points out that in many cases innovations in the petroleum refining industry came from people outside the industry but often closely related to it. Identification of internal production needs is an important task in a company's development of process technology, but so is discovering the means to fulfil those needs.

Phase 2 — Process development work in laboratories, pilot plants and production plants

This problem-solving phase is where actual systematic testing is carried out, comprising all kinds of tests from laboratory experiments through full-scale production plant trials, often including a number of different sub-processes and iterative development loops. This phase is similar to the aforementioned traditional model in Figure 8.2 for process development. As such, this traditional model could be used as a starting-point for the development of a more detailed company-specific process for this phase of the process development work process. The second phase of the process development work process is where the real development work takes place in the laboratory and pilot plant. Referring to Chapter 6, there are three options for acquiring new process technology: develop new technology within the internal organization (which this study focuses on); acquire new technology from external organizations; or a combination of both in a joint development.

The second option, which can be characterized as "open process innovation", shortcuts the total development process by using external sources, and covers two essentially different cases: the new technology already exists and can be acquired by licensing or purchase, or it must be developed within the external organization. This type of external technology transfer has been widely studied, see Lee (1993) for example; it shortcuts the total process insofar as the development phase is non-existent or small. This will be further explored in Chapter 12. The first option is when the new or improved process technology is developed mainly in-house.

Phase 3 — Transferring development results to production

This is the transfer of the results from the previous process development phase into the production environment. This phase may also have been overlooked or neglected in the past, yet it represents the part of the development process that sometimes makes the difference between success and failure. It may be improved by using cross-functional teams in the process development phase, or active participation of R&D personnel during start up (see Chapter 7).

For internal transfer of technology, which is far less studied than external transfer, the technology transfer phase is the important link between new process technology and production, just as development of new products is of little use without a successful product launch on the market. The results of a study of developer-user interaction and user satisfaction in internal technology transfer show a rather complex picture of how to obtain a smooth transfer process (Leonard-Barton and Sinha, 1993). For both types of technology transfer there may be barriers of different kinds that prevent an efficient transfer. Those barriers are not only based on and related to the transferability of the technology, but may often be of a more cultural or social character and related to communication between the supplier and the customer. The difficulties associated with technology transfer in the petroleum industry have been related to the discrepancy between the innovative individuals, often with a good theoretical background, and the production people focusing on proven trouble-free technology (Knight, 1984). The personal aspect of technology transfer is also highlighted in a paper by Langrish (1971), where the conclusion is that the most efficient transfer process is the transfer of the person with the specific knowledge.

Since a large number of the respondents in this European research project considered a formal work process for process development an important issue, the development or improvement of the process development work process should be high on the agenda of an industrial R&D organization. The fact that only 44 percent of all companies in this study already had such a work process is an indication that the development of such a process in many cases will have to start from scratch. In the development of such a company-specific work process for process development, the conceptual model presented can be a good starting-point and serve as some sort of structural framework. The conceptual model should then be further developed into a more detailed formal work process, using the previously presented definitions and language for work processes (sub-processes, process interface, process owners, etc.)

Lessons should also be learned from research about the product development work process, insofar as the work process should probably be flexible and adaptable to different types of process development and to individual company characteristics.

This exploratory study has highlighted the industrial importance of this area, and presented a new model which might be a starting-point for the further development of the process development work process. Many important questions need to be answered in this rather under-researched area:

- What are the building blocks of a more detailed process development process? How should they be put together and how should the iterative loops between them be designed?
- How much does the process development work process differ for different types of process development; from optimization (incremental development) to the development of breakthrough (radical) technology?
- How company-specific is the process development work process, and how specific is the process for different sectors of the process industries?
- How should the product development work process and the process development work process interact under different conditions?
- How should effective cross-functional teams for process development be composed?

The above questions will be discussed, and some of them also answered, in the next chapter.

8.3 Work process integrations and improvements

Product and process development is sometimes looked upon as the same activity in the process industries. It can be argued that it is not necessary to distinguish the work processes for product development and process development because product development in the process industries also partly takes the form of process development work in a laboratory.

Product and process development: two separate but interacting work processes

The strongest argument against the above point of view is that both processes start with different customers and end up with different customers, which is further illustrated in Figure 8.4. Pure process development, using the definition in Chapter 3 together with the new conceptual model, is very much an internal affair in that the process

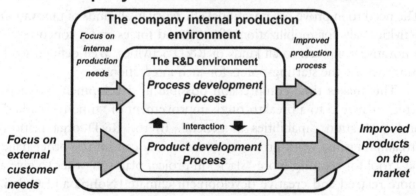

Figure 8.4   A simplified model of the product development work process and the process development work process in the process industries. The large shaded arrows symbolize that the two processes start with different customers and end up with different customers. The black arrows indicate an interaction between product and process development (after Lager, 2000).

starts and ends within the company. There is no need to interact with the product development work process. Product development and the product development work process will on the other hand include development activities that to a considerable extent are of a process technology development nature, in order to produce new products.

The importance of an interaction between product and process development has been stressed by Etienne (1981). To achieve this, an interaction between the product development work process and the process development work process must be observed. Sometimes both product and process development are included in a development project. In that case, the work processes for product and process development should interact, although different objectives and customers for the individual and separate work processes should be observed. Transfer of the process development results are however always to production, whether driven by internal production or product development needs.

Streamline innovation work processes and compress the time frame by concurrent innovation

The need to improve organizational efficiency in all kinds of innovation activities calls for mobilization of combined forces and a concurrently innovative approach to all kinds of R&D activities. Well-defined work processes are the starting-points for such integration.

This makes time compression possible in development projects, which may lead to a breakthrough improvement of "time-to-market" and innovation capabilities in general. In the R&D organization, streamlining the work processes concurrently can on the other hand be a tool for creating more "slack" in project plans, in order to give a more relaxed and creative development climate (Nohria and Gulati, 1995).

In Figure 8.5 the necessary integration of different kinds of innovation and innovation-related activities in a product development project are symbolized to illustrate the different capabilities often needed in such a project.

The activity here called "development with equipment suppliers", which is normally included in the process development activity, is also

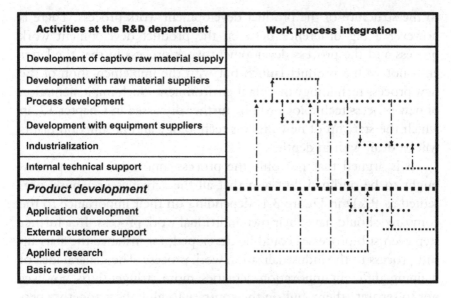

Activities at the R&D department	Work process integration
Development of captive raw material supply	
Development with raw material suppliers	
Process development	
Development with equipment suppliers	
Industrialization	
Internal technical support	
Product development	
Application development	
External customer support	
Applied research	
Basic research	

Figure 8.5   A symbolic representation of the interaction between different kinds of innovation and innovation-related activities in a product development project.

highlighted to illustrate the necessity of external collaborations in some projects. This area is however presented in much more detail in Chapter 10.

8.4 Summing-up and some issues to reflect upon

Work processes and their development are presented in this chapter, and the initial question raised is how much structure is needed in innovation. It is claimed that carefully crafted and continually improved innovation work processes are excellent tools for organizational learning and are thus important competitive tools.

A process innovation (development) work process has been structured, and it is recognized that a less formalized work process is sometimes in fact only a part of the full work process in the process industries. The process development environments that are discussed in companies lack the important idea-generation and technology-transfer phases. This lack has been corrected in the proposed process innovation work process, which as a result acquires some resemblance to the structure of the product development work process. There is however a big difference between the product development work process and the process development work process in that the latter ends not with a product launch but with the implementation of the new process technology in a plant environment. Such implementation of new process technology will be further discussed in Chapter 12, in which the start up of new process technology and production plants will be dealt with in depth.

It is argued that not only the process innovation work process ought to be outlined, but also that all the activities previously presented as R&D in Figure 3.1, depending on their importance to the company, should have their own individual work process. In a further step even sub-processes could be developed, for instance the start up sub-process to the industrialization work process. The general idea of defining different innovation activities more stringently is however not to separate them, but on the contrary to glue them together better and to provide better structures for integration. It is thus finally concluded that good work processes are interesting blueprints for corporate best practice in innovation.

In order to be "truly" effective and efficient, it is finally argued that a proper balance of well designed structural organizations and appropriate complementary formal work processes has the potential to deliver future lean process innovation.

- *Does your company use formal work processes to give guidance for different kinds of work in the R&D organization, or is it on the contrary believed that such formal presentations are only tools for innovation bureaucrats?*
- *If you already use formal work processes in your R&D, how well is the ownership defined? How well are interfaces presented? What does the content look like, and how efficient are the processes?*
- *If you have different kinds of R&D work processes, how well are they integrated?*
- *Does your firm already have a formal work process for process innovation or if not, is the structural model applicable as a starting point in your firm?*

- *Starting with the new structural model presented in Fig. 8.3, benchmark the importance of each phase and consider how good your firm process innovation is during those different phases!*
- *The formal structure of a process innovation work process has been outlined with its interfaces, input and output. What has not been further discussed in detail, however, is the question of how to secure a proper input to such a work process from a process innovation strategy or from complementary new idea generation. How could this be arranged in your company, and what is your firm's approach to idea management in innovation?*
- *In the perspective of lean innovation, do you believe that there is a potential to improve the efficiency of your firm's (informal or formal) process innovation work process?*

Chapter 9

The Development of a Process Innovation Work Process — An LKAB Case Study

"As a process engineer, you would most certainly be of the opinion that it is impossible to control or improve a technological process that lacks a proper flowsheet. As an analogue to this statement you should ask the question: 'How do we control our innovation work processes, and what do their flowsheets look like?'"

Thomas Lager, 2008

There are fundamentally two different roads to follow in the improvement of work processes, although a combination of both may be the optimal solution. The classical continual improvement road in the framework of corporate Total Quality Management (TQM) initiatives is something that is always important to consider (Bergman and Klefsjö, 1994). At the beginning of the 1990s, Michael Hammer challenged this approach with his new re-engineering concept under the banner "Don't automate, obliterate!" (Hammer, 1990).

Further research by Cooper (2008) has gradually refined his stage gate work process for product development presented in the previous chapter. A late review gives many important suggestions for the next-generation stage gate work process, such as: an adaptable process and an efficient, lean and rapid system; scaling to suit different risk-level projects; a flexible process; more effective governance and the use of scorecards to make better go/kill decisions; employing success criteria at gates and displays of in-process measurements at gates, etc. But what about work processes for process innovation?

In most companies today the need for a product innovation work process goes without saying. The importance of process innovation in many sectors of the process industries should however make the development of a process innovation work process a growing concern in many firms. This case study, which contains selected parts from a previous publication (Lager *et al.*, 2010), describes the development of such a work process at LKAB, a Swedish producer of high-quality iron ore pellets mainly for the European steel industry.

The complete work process is presented, including a cross-functional process map, checklists and a supplementary process description. The development of the work processes was a combination of a recording of LKAB's best practice, the use of externally published research results on work process development and a minor re-engineering effort. Finally, experience from development and organizational implementation is discussed.

9.1 Developing the LKAB process innovation work process

LKAB is an international high-tech minerals group that produces iron ore products for the steel industry and other mineral products for various industrial applications (www.lkab.com, 2009). The main products in 2008 were iron ore pellets for the production of iron in blast furnaces (13.6 million tonnes) and pellets for direct reduction processes (4.3 million tonnes). The LKAB Group has about 4,000 employees and LKAB's iron ore products are sold mainly to customers in Northern Europe, North Africa and the Middle East.

Research & development work at LKAB is organized as part of the technology & business development business unit, which also carries out investment projects. The organization, with around 180 employees, is headed by the vice president of technology & business development directly under the CEO. Research & development activities are organized in three departments: mining technology,

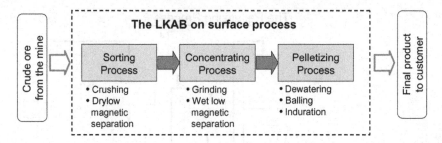

Figure 9.1   The LKAB production process at the Malmberget site. The mining level is situated 1,000 metres below ground level, and the crude ore is transported up to the surface. The ore is ground and upgraded in the sorting and concentrating plants and further pumped to the pellet plants. The processes operate on a continuous year-round basis, 24 hours per day. The pelletizing process with its unit operation processes is further presented in Fig. 9.2.

process technology and metallurgy (recently transferred to sales & marketing). Mining technology is responsible for R&D work underground, process technology is responsible for R&D work for on-surface processes and internal production support, while metallurgy is responsible for product development and technical support to external customers.

The share of total R&D resources devoted to process technology was approximately 35 percent during 2008. During the recent years of heavy investments, the activities of process technology have shifted to some extent from process development (process innovation) to production support and support to industrialization. The process innovation work process is thus a work process affiliated to the process technology department for the development of new or improved process technology (on-surface production processes).

The development of a process innovation work process was launched on the initiative of the manager of process innovation at LKAB. At that time a product innovation process had already been developed within the company, and the outcome of that work was considered of such interest that an attempt to develop a similar and more formal work process for process innovation was deemed of interest. A small team with the necessary deep knowledge in this area was created by the manager to develop of the process. The development

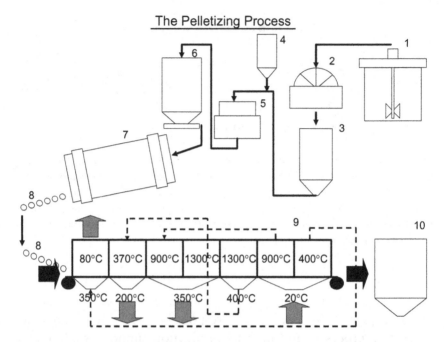

Figure 9.2   The pelletizing process in Malmberget. The iron ore slurry is dewatered and mixed with a binder. In the next step the feed is balled and screened to green pellets with a diameter of 10 mm. The pellets are sintered and cooled in the induration furnace (straight grate).

work was initially of a very iterative kind, but gradually arrived at a list of the structural parts which such a work process should contain. However, the development strategy rested on a number of cornerstones that the working group agreed upon right at the start of the project.

An organizational framework and up-front ideas

From the outset the ambition was to try to combine a TQM-like work model (Bergman and Klefsjö, 1994) with a more radical re-engineering model (Hammer, 1990). The group developing the work process was of the opinion that process innovation work at LKAB was already in many respects a successful undertaking, and it was decided to try first of all to capture this working mode as well as possible, including more obvious minor improvements. Nevertheless, the lessons learned from previous research in innovation and technology management were not to be neglected, and experience from successful product development work processes was not to be ignored. Four areas were identified as important cornerstones for more radically improved development behaviour:

A stage gate approach

Cutting up an innovation project into smaller chunks, keeping the innovation investment low in the early phases, and gradually increasing the investments when more knowledge has been gained is a model proposed by Cooper for product development projects (Cooper, 1988b). A similar approach was selected in the development of a process innovation work process, and the adoption of a stage gate philosophy was agreed upon at the start up of the work process development project.

More focus on pre-project activities — the fuzzy front end

In view of research results on idea generation and evaluation (Cooper, 1988a; Verworn *et al.*, 2008), and on the importance of improving the idea generation process in industrial innovation and product development in particular, one can suspect that improved idea generation in process innovation is important too. Acknowledging that there is no "equality" in a well-functioning idea generation process, and with the intention of letting only the ideas with the strongest potential survive, the fuzzy front end of the process innovation work process needs to be emphasized.

Stronger focus on internal and external cross-functional collaboration

All kinds of R&D depend nowadays on various forms of collaborative approaches (Chesbrough and Schwartz, 2007), including not only classic cross-functional in-house collaboration but external collaboration with different partners and using different forms of collaboration. It is now an established truth that such cross-functional collaborative behaviour not only reduces the lead time but enhances the overall efficiency of development work. Previous research on process innovation supports this belief (Lager and Hörte, 2002; Pisano, 1997), and in the light of LKAB's process innovation experience to date it was decided to underline such collaborative behaviour, with special emphasis on early cross-functional development with production.

Improved technology transfer of R&D results

Technology transfer has long been recognized as a weak area in industrial R&D (Holden and Konishi, 1996; Leonard-Barton and Sinha, 1993; Levin, 1993). Research and development results that do not reach the "customer" are seriously wasted resources. Because of that an improved dialogue with the customer (in this case often production) was included in the development of the work process right from the outset. This intended technology transfer could be described as a more gradual information sharing and participation, not just a handover.

From simplified structural process map to a cross-functional process map and checklists

The work of drawing a structural process map began by defining the limits of the work process and its input and output. The next step was to identify necessary "gates" in the work process, after which a first preliminary structural process map could be drawn. It was learned that a good structural process map evolves in the course of a number of meetings, and the original version needed to be revised a number of times. The different phases were identified, named, and further structured into sub-phases when necessary.

Finding the optimum, and more firm-generic, division of a process into phases is not easy, so enough time was allowed, as this was considered one of the most important operations in the formulation of the work process. This first type of map was called a structural process map, in which the selected gates and decision points were marked out with supplementary information about the decision groups and also sometimes additional advisory groups. To assist in this work, it was discovered that a good way to start was by giving an outline description of a previous project to get a hands-on feeling for how this type of work had been structured in the past and how the contents of the various phases had previously been described. Having established a preliminary structural map, the associated checklists were developed in an iterative approach with the development of a cross-functional process map.

The checklists were linked to the previously identified gates, and lists were made of the criteria that must be satisfied to pass through the relevant gate, or items to provide at other checkpoints. The contents of the checklists were usually parts of prior practice in task and project management, supplemented by important new points that were identified as work progressed. Out of all the checklists for the process innovation work process, Table 9.1 illustrates the one for passing the gate between pre-studies and the

Table 9.1 A Partly Disguised and Simplified Checklist for Starting the Process Development Phase in the Process Innovation Work Process.

---

1. Describe the new process that is intended to be developed
    1.1 Production benefits
    1.2 Describe the new process (as well as is possible now)
    1.3 Process understanding (how/why do we believe it works as stated?)
2. Calculations
    2.1 Tentative investment needs
    2.2 Tentative operating cost
    2.3 Potential and risk
3. Influences on implementation (positive & negative)
    3.1 Production influence?
    3.2 Product influence?
    3.3 Environment (internal & external)?
4. Securing internal management support
5. Project plan
6. Project collaboration
7. When necessary, negotiations with union representatives

---

project development phase. This is of particular interest, since it illustrates well the complexity of process development projects, and it was as such by far the longest checklist.

A review and critical examination of existing methods, procedures and inherent activities in the total work process took place parallel with the development of the checklists. Making this kind of cross-functional process map is, as previously described, a matter of identifying internal and external interactions during the work process.

As with the structural process map, the production of the cross-functional map was to some extent a process of maturation through an ongoing discussion; this was found preferable to a brief concentrated effort. A simplified and partly disguised version of the cross-functional process map for process innovation is presented in Figure 9.3. The collaborative partners (here disguised) were listed along the vertical axis and the collaboration intensity was indicated by different shades of grey (the darker, the stronger). The cross-functional process map outlines with whom to collaborate during a process innovation project, when such collaboration should take place, and how further such collaboration could be organized. Apart from illustrating the need for more general internal and external collaboration, the importance of collaboration between R&D and production was emphasized. Early integration is desirable from the start in order to secure good participation and future acceptance of new process technology in the implementation phase.

9.2 Integrating all information into a "mini-guide"

During the development of the work process it was realized that there was a need for a supplementary descriptive document, as the very large volume of information built into the process maps and checklists might otherwise be lost to the users. At LKAB such a document is called a work process description, and together with the process maps and checklists it makes up the total "mini-guide" for each individual work process. The process description, presented in the following section in a condensed version, is intended to be a synthesis of all the documents

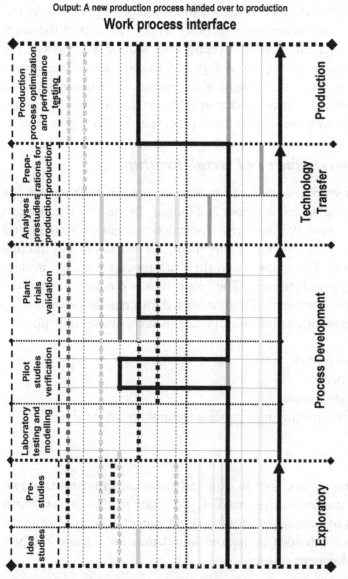

Figure 9.3 A simplified cross-functional process map. The texts in the map showing collaborative partners have been disguised. The dark black line represents a strong commitment and is often the "critical line". The lines with lighter shades of grey represent less intensive activities of collaborative partners.

for the work process and links the process maps to the checklists. This work process description is intended to be compiled gradually by the individual process owner and supervisor as the last step of the work process development. Maybe covering only around 10 pages at the outset, the process description is expected to grow into a more comprehensive document in the light of experience from using the work process. In this case a substantial part of the true "mini-guide" for the users of the process innovation work process at LKAB is presented.

Work process interfaces and overall structure

Like all other work processes, the process innovation work process has its interfaces and its defined input and output. In this case the input to the process consists of long-term production planning, a strategy and plan for product & process development, applied R&D and internal idea generation. The output from the work process is a new production process up and running. The task is thus not complete until the results have been transferred and translated into regular production.

The process innovation work process is divided into four phases:

- Exploratory work
- Development
- Technology transfer
- Production (putting new process technology on stream)

Exploratory work — Phase 1

This preliminary phase is of an exploratory nature and consists of two stages, idea studies and pre-studies. The idea studies are an initial investigation of whether an idea has development potential, while the pre-studies are intended to lay the foundation for a more comprehensive development project.

Idea studies

It is LKAB's policy that getting approval to start an idea study should be a simple matter involving a minimum of bureaucracy, so authority for approval rests with the immediate superior. Problem analysis and a review of what has previously been done in the field are important components of the idea study. Preliminary experiments on a laboratory scale to test the idea, and discussions with experienced production people provide further input. The aim of the idea study is to determine whether an idea is feasible or not. If the results of the idea study are interesting enough to justify further investigation, the next step is to draw up a proposal for a pre-study.

Pre-studies (development)

The proposal for a pre-study should begin by presenting the results of the idea study and should then put forward a simple plan for the pre-study and secure acceptance for it in the organization. The decision to proceed is taken by the departmental management group. A broader situational analysis, continued laboratory studies and initial modelling (simulation) are important elements in a pre-study. Issues which the pre-study also ought to clarify include working environment, energy, environmental impact, and whether the proposed project could affect the product in any way. A tentative cost-benefit analysis is also an important result of the pre-study. If the pre-study indicates that the idea is worth pursuing further, the next step, to be decided on by the departmental management group, is to draw up a proposal for a process development project as a basis for starting the next phase.

Process development — Phase 2

The development phase is characterized by greater plannability and structure compared to the exploratory phase. An important success factor, in LKAB's experience, is a well-defined allocation of resources to the project. The threshold for initiating a process development project should be significantly higher than for idea studies and pre-studies. The work done in the exploratory phase should now have provided a firmer basis for formulating a detailed project proposal. The decision to start the project is taken by the technology unit management.

Process development work is characterized by a high degree of complexity, and several areas must be elucidated before the project starts (see previously presented checklist). The development phase is further divided into three stages:

- Laboratory scale studies (modelling)
- Pilot scale studies (verification)
- Production scale studies, demonstration (validation)

Because production is the main customer for process development work, collaboration obviously needs to be set up and representatives of production involved at the earliest possible stage. A good cross-functional team must be assembled according to the needs and nature of the project.

Laboratory testing and modelling

Laboratory studies provide an opportunity to try out a large number of different kinds of experiments, to gain a basic understanding of unit operational phenomena, and thereby to construct conceivable process layouts. A further characteristic feature of laboratory studies is that they are repeatable and that the experimental environment is easy to control. At this stage, the laboratory studies can be very extensive, and the aim should be to do as much of the development work as possible here, which is much more cost-effective than doing it later during pilot scale studies. The continued deepening and broadening of the laboratory work thus aims at establishing a firmer basis with fewer options before the next phase of pilot scale studies begins. An important success factor here is close collaboration between researchers, engineers and laboratory staff on planning, conducting and evaluating the experiments.

Pilot studies — verification

Proceeding from laboratory to pilot scale is an important step towards gaining better understanding of the process. Pilot studies facilitate scaling-up and offer opportunities to study complete circuits with circulating loads. This stage also enables production of larger quantities of products for further testing and experimentation. The large increase in costs associated with pilot scale studies necessitates striking a delicate balance, but even so, pilot studies still allow more freedom and involve less risk than production-scale studies. Pilot studies require more detailed planning and greater resources compared to laboratory studies. The time needed for evaluation and reporting is also longer, and is often underestimated. Here too, future users of the process should be involved in the trials to familiarize them with the process and facilitate later technology transfer.

Plant trials — validation

Full-scale trials are usually conducted at a production plant. Their purpose is to validate design solutions and control systems. Extended trials also give an opportunity to study wear and corrosion, and to make larger quantities of the product. The following issues must be clarified before the trials start:

- Risk analysis of project implementation.
- Risk analysis with regard to impact on the product, the environment and the process.
- Detailed test plan.

Full-scale trials are normally conducted in a production plant. This makes heavy demands on co-ordination with regular operations, and likewise on the process control and regulating systems. In terms of time, the initial trials are often run batchwise, but with the ultimate object of achieving continuous operation. The trials also generate larger quantities of products (hundreds of tonnes per hour) which may have to be handled separately for use in further trials by external customers. Important success factors are cross-functional trial planning, the presence of supervisory personnel on-site day and night during the whole trial period, and early and intimate collaboration between production, maintenance and R&D. The results of the trials are intended to indicate which of the following three options should be chosen:

- Negative result: Final report and termination of project.
- Inconclusive but interesting result: Some kind of iterative restart.
- Clear result and objective achieved: Proceed to implementation in production.

If the results involve major investment in new equipment, the project moves on to the construction work process. If the new or improved process technology does not require any changes in plant equipment or only minor installation and construction work, the project moves to the next phase of the work process.

Technology transfer — Phase 3A

The object of this phase is to bring the results of experimentation and trials into the existing production process smoothly and without detriment to quality. Experience shows that even minor process development projects can result in serious disruption of production if the implementation in production is not handled properly. Success factors are that the future user has been actively involved in earlier phases of the project, and that the project organization is there to support the production organization even after handover. This phase is very much a matter of transferring knowledge from the developers to the production people. It comprises:

1. Background reports: Documentation containing process concepts and results from previous trials provides the background to and basis of technology transfer.
2. Operating instructions: One important item of knowledge emerging from the trials is how the process should be run. This knowledge must be translated into an easy-to-follow operators' manual.
3. Training: Various types of training courses will be necessary to assure transfer of knowledge and promote confidence in the new process on the part of its future operators.

When this has been done, the process is handed over to production.

Production — Phase 3B

The handover takes place in two stages. In the first stage, production assumes responsibility for operation, while responsibility for the process itself remains with the project group (production takeover). The next stage is final process takeover. The purpose of these two distinct checkpoints in the work process is to emphasize that the process developers' responsibility does not end when the process goes into operation. They continue, in collaboration with production, to fine-tune the process right up to the formal handover. Items to be dealt with before this takes place are:

1. Performance test.
2. Fulfilment of objectives: If criteria are not satisfied, a checklist of outstanding items.
3. Final report and termination of project.

The new or improved process technology is now formally taken over by production. That concludes the process development project.

9.3 Lessons learned and expected outcomes

Some work processes developed at LKAB have not yet been used or tested operationally to any great extent by their respective organizations; this is also the case for the process innovation work process. Once a work process has been developed, it must not only be introduced in the organization concerned but operated over an extended period of time before it is evaluated. The following is a summary of experience to date with the development and implementation of the process innovation work process.

Lessons learned

The formalized structure, with its process maps and checklists, can be regarded partly as a description of what, at best, is already being done at LKAB in the course of a very well run project. Good, smoothly operating work processes are thus likely to have grown out of current practice that has been more clearly structured and where the content has been improved with new useful information. Nevertheless, experience at LKAB shows that it takes a while for individual employees to adopt a work-process mindset, so the new way of doing things must be allowed time to mature within the organization. One must respect the fact that not everybody in the organization welcomes a new formal work process with open arms.

In the selection of a project team, the members must all have a strong commitment to the cause and must be deeply knowledgeable in the area, with first-hand practical development experience. It is not only necessary to select and involve the process owner from the start, but also to include an internal supervisor with a keen interest and time available for implementation and further improvements. Prior knowledge in the general area of innovation management was felt to be an important asset, especially good knowledge in the area of work processes. If this experience is not available in-house, an outside facilitator is recommended. The project must be carefully planned, and necessary resources must be allocated (try to avoid a crash project).

Start to think from the outset about the organizational implementation and the administration of the work process, e.g. data registration and overall management. During the development it was found that the need to solve specific problem areas was an important driver in promoting improved innovation behaviour and more efficient work practices. As a consequence of that it was afterwards thought that a "pilot test" of the work process on a couple of real projects would have been a useful exercise. After the development was completed it was noticed that not enough time had been allocated to sorting out how the process should be managed, e.g. how the different gates should be operated, and how to secure proper input of ideas and development strategies to the process. At the implementation stage it was learned, the hard way, how best to proceed. At LKAB it was found that an initial focus on more frequent users was to be recommended after an introduction to section heads and specialists. Hands-on work with the charts and checklists proved more valuable than lecturing.

An overall co-ordinator (a super-user) was assigned from the start to all work processes, which proved to be a well-functioning solution and probably an important step in ensuring that a work process will be kept alive and not only be brought forward during quality audits. It was found that the development of a formal work process is something that needs management support, and as such should be regarded as an important tool for the company's improvement of management of innovation & technology.

It should be noted that the intention in describing and specifying individual work processes is not to obstruct or limit interaction. On the contrary, it was observed that well-formulated work processes make it easier to see how interaction with adjacent processes could be established and improved. If the work or the pre-study phase proves to involve machine development, that part of the work on the project will conform to and follow the equipment development work process. Similarly, if the newly developed process requires process control, that part of the work will conform to the process control work process. If, when development has been completed, the new process calls for new construction on a major scale, the project becomes a construction project and becomes input to the construction work process. If, on the other hand, a new or improved product needs new or improved process technology, work on that should of course be initiated as soon as possible, and should be guided by the process innovation work process. The objectives from the product innovation work process become the input, and consequently such work is to be considered as product development and a part of the product development project.

Expected outcomes

Substantial benefits are foreseen at LKAB from using a process innovation work process. Although a certain time constant can be expected with regard to improved development results from its future use, the overriding aim is more efficient process innovation. It is hoped that improvements in efficiency will be gained partly from a better inflow of ideas and pre-studies that generate good and important development projects (better input to the work process) and partly from a better transfer of results (better output from the work process). There are many other aspects and areas where a well-formulated work process is expected to benefit LKAB; some of them are listed below, not in ranking order:

- Better quality in project management and shorter lead times as a result of better input and smoother transfer of results.
- Lower risk in development projects because decisions are taken in well-founded successive steps.
- Generally better project management because the work process gives guidance and a general description of how development work ought to be tackled (no need to re-invent the wheel for each new project).
- Better-structured project plans to supplement the company's project management instructions (especially useful to inexperienced project managers).

9.4 Summing-up and some issues to reflect upon

The growing importance of process innovation in many sectors of the process industries should make the development of a process innovation work process a growing concern. The process innovation work process presented in this chapter can serve as an embryo and a starting-point for the development of other companies taking a similar approach to process innovation.

The building blocks in a process innovation work process have been identified, and this process can be used for both incremental and radical process development. It was the opinion of the project group that the layout of the work process is probably fairly generic and that the overall structure could also be used as a starting point for the development of similar work processes in other sectors of the process industries.

During the development of the work process, a common language and way of thinking were created, which in itself was considered as one important outcome. The building blocks of such a process were identified, and the work process description using a cross-functional process map, checklists and process description is introduced as a useful element. The importance of selecting a skilled and knowledgeable team for this kind of development, and of considering the implementation at the outset, is emphasized.

The lessons learned in the development of this work process provide practical guidance for other companies on how to organize such an endeavour and how to avoid some unnecessary pitfalls on the way. The final conclusion of the project group was that if properly applied, this work process will give structure, methodological guidance and efficiency in process innovation; but reaping those fruits will call for persistence and discipline at all levels in the company.

- *A well designed work process is claimed in Chapter 7 to be an excellent tool for organizational learning. After reading this case study, do you find that this is a likely outcome or do you believe that this concept is just another empty management "buzzword"?*
- *Does your firm give any guidance similar to the LKAB mini-guide to new managers of process innovation projects on how to plan such a project and how to go ahead?*
- *Examine the overall structure of the process innovation work process presented in Fig. 9.3. Do you believe that the phases could be similar for your kind of process innovation, or in what respects do you believe that your work process should be structured differently?*
- *The LKAB work process has been presented in a mini-guide including a process description, checklists and process maps. Do you think that this is functional, or do you find that a different presentation is needed in your company or that there may be better ways to do it?*
- *The organization of the development and implementation of the work process in LKAB was a multi-stage procedure. Reflecting on the pros and cons presented in this case study, how would you organize and carry out a similar development and implementation of a process innovation work process in your company?*

References

AT&T (1988). *Process Quality Management & Improvement Guidelines.*

Bergfors, M (2007). Designing R&D organisations in process industry. Department of Business Administration and Social Sciences. Luleå, Luleå University of Technology.

Bergfors, ME and Larsson, A (2009). Product and process innovation in process industry: a new perspective on development. *Journal of Strategy and Management*, 2, 261–276.

Bergman, B and Klefsjö, B (1994). *Quality: from Customer Needs to Customer Satisfaction.* Lund, Studentlitteratur.

Booz Allen & Hamilton (1982). New Product Management of the 1980s.

Bower, DJ and Keogh, W (1996). Changing patterns of innovation in a process-dominated industry. *International Journal of Technology Management*, 12, 209–220.

Bowonder, B, Miyake, T and Butler, J (1995). Creating and nurturing virtual organisations: strategic trajectory of Toshiba. *R&D Management.* Miniato; Italy.

Brockhoff, KK, Koch, G and Pearson, AW (1997). Business process re-engineering: experiences in R&D. *Technology Analysis & Strategic Management*, 12, 209–220.

Cardinal, LB (2001). Technological innovation in the pharmaceutical industry: The use of organizational control in managing research and development. *Organization Science*, 12, 19–36.

Chandler, AD (1962). *Strategy and Structure: Chapters in the History of the American Industrial Enterprise.* Cambridge, MA: MIT Press.

Chesbrough, H and Schwartz, K (2007). Innovating business models with co-development partnerships. *Research Technology Management*, January–February, 55–59.

Chiesa, V (2001). *R&D Strategy and Organisation: Managing Technical Change in Dynamic Contexts.* London: Imperial College Press.

Chronéer, D (2003). Have process industries shifted their centre of gravity during the 90's? *International Journal of Innovation Management*, 7, 95–129.

Clark, KB and Fujimoto, T (1991). *Product Development Performance: Strategy, Organization, and Management in the World Auto Industry.* Boston, MA: Harvard Business School Press.

Cleland, DI and Kerzner, H (1986). *Engineering Team Management.* New York: Van Nordstrand Reinhold Company Inc.

Contreras, J, Sistemas, I and Ferrer, JM (2005). Dynamic simulation: a case study. *Hydrocarbon Engineering*, May.

Cooper, RG (1988a). Predevelopment activities determine new product success. *Industrial Marketing Management*, 17, 237–247.

Cooper, RG (1988b). *Winning at New Products*. London: Kogan Page.

Cooper, RG (1994). Perspective. Third-generation new product processes. *Journal of Product Innovation Management*, 11, 3–14.

Cooper, RG (2008). Perspective: The stage-gate idea-to-launch process — update, what's new, and nexgen systems. *Journal of Product Innovation Management*, 25, 213–232.

Cooper, RG and Kleinschmidt, EJ (1986). An investigation into the new product process: steps, deficiencies and impact. *Journal of Product Innovation Management*, 3, 71–85.

Damanpour, F (1996). Organizational complexity and innovation: Developing and testing multiple contingency models. *Management Science*, 42, 693–716.

Enos, LJ (1962). *Petroleum Progress and Profits*. Cambridge, MA: MIT Press.

Etienne, CE (1981). Interactions between product R&D and process technology. *Research Management*, 24(1), 22–27.

Ettlie, JE, Bridges, W and O'Keefe, R (1984). Organization strategy and structural differences for radical versus incremental innovation. *Management Science*, 30, 682–695.

Freeman, C (1990). Technical Innovation in the world chemical industry and changes of techno-economic paradigm. In *New Explorations in the Economics of Technological Change*, C Freeman, and L Soete (eds.), pp. 74–91. London: Pinter Publishers.

Hall, RH (2002). *Organizations — Structures, Processes and Outcomes*. Upper Saddle River, NJ: Prentice Hall.

Hammer, M (1990). Reengineering work: don't automate, obliterate. *Harvard Business Review*, 104–112.

Hammer, M (2007). The process audit. *Harvard Business Review*, 1–14.

Harding, JA and Popplewell, K (2000). Simulation: an application of factory design process methodology. *Journal of Operational Research Society*, 51(4), 440–448.

Harryson, SJ (1997). How Canon and Sony Drive product innovation through networking and application-focused R&D. *J Prod Innov Manag*, 14, 288–295.

Hayes, RH, Pisano, GP and Upton, DM (1996). *Strategic operations: Competing through capabilities*. New York, NY: The Free Press.

Heygate, R (1996). Why are we bungling process innovation. *The McKinsey Quarterly*, 2, 130–141.

Holden, PD and Konishi, F (1996). Technology transfer practice in Japanese corporations: meeting new service requirements. *Technology Transfer*, Spring–Summer, 43–53.

Hutcheson, P, Pearson, AW and Ball, DF (1995). Innovation in process plant: a case study of ethylene. *Journal of Product Innovation Management*, 12, 415–430.

Hutcheson, P, Pearson, AW and Ball, DF (1996). Sources of technical innovation in the network of companies providing chemical process plant and equipment. *Research Policy*, 25, 25–41.

Hörte, S-Å (1997). Social network analysis and manipulative management. Luleå University of Technology, Department of Business Administration and Social Sciences.

Johnson, G and Scholes, K (1999). *Exploring Corporate Strategy*. London: Prentice Hall Europe.

Kahn, KB (1996). Interdepartmental integration: A definition with implications for product development performance. *Journal of Product Innovation Management*, 13, 137–151.

Kline, SJ (1985). Innovation is not a linear process. *Research Management*, July–August, 36–45.

Knight, KE (1984). Technology transfer in the petroleum industry. *Journal of Technology Transfer*, 8, 27–34.

Kotter, JP (2007). Leading change — why transformation efforts fail. *Harvard Business Review*, January.

Kuemmerle, W (1997). Building effective R&D capabilities abroad. *Harvard Business Review*, March–April, 61–70.

Laganier, F (1996). Dynamic process simulation — trends and perspectives in an industrial context. *Computers and Chemical Engineering*, 20, 1595–1600.

Lager, T (2000). A new conceptual model for the development of process technology in process industry. *International Journal of Innovation Management*, 4, 319–346.

Lager, T (2002). A structural analysis of process development in process industry — A new classification system for strategic project selection and portfolio balancing. *R&D Management*, 32, 87–95.

Lager (2008). Work processes in Process Industry. Stockholm.

Lager, T, Hallberg, D and Eriksson, P. Developing a process innovation work process: The LKAB experience. To appear in *International Journal of Innovation Management*.

Lager, T and Hörte, S-Å (2002). Success factors for improvement and innovation of process technology in process industry. *Integrated Manufacturing Systems*, 13, 158–164.

Langrish, J (1971). Technology transfer: Some British data. *R&D Management*, 1, 133–135.

Larson, EW and Gobeli, DH (1988). Organizing for product development projects. *J Prod Innov Manag*, 5, 180–190.

Latour, B and Woolgar, S (1986). *Laboratory Life — The Construction of Scientific Facts*. Princeton, NJ: Princeton University Press.

Lawrence, PR and Lorsch, JW (1967). *Organization and environment: Managing differentiation and integration*. Boston, MA: Harvard Business School Press.

Lee, G (1993). Closing the performance gap through technology transfer: linking theory and practice. *Int. J. Technology Management, Special Issue on Manufacturing Technology: Diffusion, Implementation and Management*, 8, 236–243.

Leonard-Barton, D (1992). The factory as a learning laboratory. *Sloan Management Review*, 34, 23–38.

Leonard-Barton, D and Sinha, DK (1993). Developer-user interaction and user satisfaction in internal technology transfer. *Academy of Management Journal*, 36, 1125–1139.

Levin, M (1993). Technology transfer as a learning and development process: an analysis of Norwegian programmes on technology transfer. *Technovation*, 13, 497–518.

Liebeskind, D (1998). Reengineering R&D work processes. *Research Technology Management*, 45, 43–48.

Liker, JK and Meier, D (2006). *The Toyota Way Fieldbook: A practical guide for implementing Toyota's 4Ps*. New York: McGraw-Hill.

Lim, LP, Garnsey, E and Gregory, M (2006). Product and process innovation in biopharmaceuticals: A new perspective on development. *R&D Management*, 36, 27–36.

Lorenz, C (1991). Team-based engineering at Deere. The McKinsey Quarterly, 4.

Malone, TW, Crowston, K and Herman, GA (2003). *Organizing Business Knowledge: The MIT Process Handbook*. Cambridge, MA: MIT Press.

Margherita, A, Klein, M and Elia, G (2007). Metrics-Bases Process Redesign with the MIT Process Handbook. *Knowledge and Process Management*, 14, 46–57.

Matheson, D and Matheson, J (1998). *The Smart Organization: Creating Value through Strategic R&D*. Boston, MA: Harvard Business School Press.

Melan, EH (1989). Process management: a unifying framework for improvement. *National Productivity*, 8, Autumn, 395–406.

Mintzberg, H (1999). *Structure in Fives: Designing Effective Organizations*. Englewood Cliffs, NJ: Prentice-Hall.

Moenaert, RK, Souder, WE, De Mayer, A and Deschoolmeester, D (1994). R&D — marketing integration mechanisms, communication flows, innovation success. *J Prod Innov Manag*, 11, 31–45.

Moors, EHM and Vergragt, PJ (2002). Technology choices for sustainable industrial production: Transitions in metal making. *International Journal of Innovation Management*, 6, 277–299.

Morgan, G (1986). *Images of Organization*. London: Sage Publications.

Nadler, D and Tushman, M (1997). *Competing by Design: The Power of Organizational Architecture*. New York: Oxford University Press.

Nohria, N (1995). Note on organization structure. *Harvard Business Review*, May–June.

Nohria, N and Gulati, R (1995). What is the optimum of organizational slack? A study of the relationship between slack and innovation in multinational firms. *Academy of Management Best Paper Proceedings*, 32–36.

Norling, PM (1997). Structuring and managing R&D work processes — Why bother? *Chemtech*, October, 12–16.

Palluzi, RP (1992). *Pilot Plant Design, Construction and Operation*. New York: McGraw Hill.

Palluzi, RP (1997). Succed at crash pilot-plant construction. *Chemical Engineering Progress*, 93.

Pisano, GP (1997). *The development factory: Unlocking the potential of process innovation*. Boston, MA: Harvard Business School.

Pisano, GP and Wheelwright, SC (1995). The new logic of high-tech R&D. *Harvard Business Review*, 73, 93–105.

Rice, MP, O'Connor, GC, Peters, LS and Morone, JG (1998). Managing discontinuous innovation. *Research Technology Management*, 41, 52–58.

Robbins, SP (1983). *Organization Theory: Structure, Design and applications*. Englewood Cliffs, NJ: Prentice-Hall Inc.

Sapienza, AM (1995). *Managing Scientists — Leadership Strategies in Research and Development*. New York: Wiley-Liss.

Sastry, AM (1997). Problems and paradoxes in a model of punctuated organizational change. *Administrative Science Quarterly*, 42, 237–275.

Schroeder, R, Van De Ven, A, Scudder, G and Polley, D (1986). Managing innovation and change processes: findings from the Minnesota innovation research program. *Agribusiness*, 2, 501–523.

Shelden, RA and Dunn, IJ (2001). Dynamic simulation: Modeling processes, the environment, the world. *Chemical Engineering Progress*, 97(12), 4–48.

Skinner, W (1992). The Shareholder's Delight: companies that achieve competitive advantage from process innovation. *International Journal of Technology Management*, 7(1–3), 41–48.

Taylor, WT (1916). The Principles of Scientific Management. In *Classics of Organization Theory*, JM Shafritz and JS Ott (eds.), pp. 61–72. Belmont, CA: Earl McPeek.

Tidd, J, Bessant, J and Pavitt, K (2001). *Managing Innovation. Integrating Technological, Market and Organizational Change*. Chichester, West Sussex: John Wiley & Sons Ltd.

Tidd, J and Pavitt, K (2000). Corporate versus divisional research and development. In *Technology Management Handbook*, R Dorf (ed.) Boca Raton, FL: CRC Press.

Tottie, M and Lager, T (1995). QFD — linking the customer to the product development process as a part of the TQM concept. *R&D Management*, 25(3), 257–267.

Utterback, JM (1974). Innovation and the diffusion of technology. *Science*, 183, 620–626.

Utterback, JM (1994). *Mastering the dynamics of innovation: How companies can seize opportunities in the face of technological change*. Boston, MA: Harvard Business School Press.

Verworn, B, Herstatt, C and Nagahira, A (2008). The fuzzy front end of Japanese new product development projects: impact on success and differences between incremental and radical projects. *R&D Management*, 38, 1–19.

Von Clausewitz, C (1984). *On War*. Princeton, Princeton University Press.

Woodward, J (1965). *Industrial Organizations: Theory and Practice*. London: Oxford University Press.

Young, B, Monnery, W and Svrcec, W (2001). Dynamic simulation improves gas plant. *Oil & Gas Journal*, 99, 54–57.

PART 4

OPEN PROCESS INNOVATION

In the process industries it is not so common any more for individual firms to develop and manufacture their own process technology/ equipment, which makes them dependent on external suppliers of process equipment. Historically, it can be seen that many equipment-manufacturing companies have grown from collaboration with domestic process firms to the point where they now serve customers primarily active on the global market (Auranen, 2006). The process industries, especially in the Nordic countries, have a long tradition of collaborative development between process firms and suppliers of new process technology. Such collaboration has historically produced a win-win situation where the process companies, as early users, have gained access to novel technology and equipment needed to process domestic raw materials, whereas the equipment suppliers in a geographically close and often mutually trusting relationship have gained an efficient means of testing prototypes and developing new equipment. Such collaboration between process firms and equipment suppliers is today often called "open innovation".

Another important area of open innovation in the process industries is the collaboration between process firms and their customers and end-users. Development with the external customer is not confined to the company's own product development and external customer technical services and support. In the process industries, a substantial part of the innovation and innovation-related activities of a company often lies in the area of helping the company's customers to use its products in a better way and assisting them in the improvement of their own processes and products. This area is generally designated "application development" by industry professionals. Companies in the process industries have long since identified this area of development as one of great industrial importance, but there seems to be a scarcity of published information or guidelines on how to carry out this kind of development activity in an efficient manner.

In view of the process innovation work processes presented in the two previous chapters, the final step is the implementation and start up of new process technology in the production system. Reviewing the published literature on the start up of process technology and process plants in the process industries, one gets the impression that there has been too much focus on the start up of the equipment. Too little attention has been paid to the final outcome of such a startup — a well-functioning production process and products of high quality. In measuring startup success these aspects must be more carefully evaluated; the mechanical and operational aspects of the equipment startup should come afterwards!

Chapter 10

Process Innovation with Equipment Manufacturers — a Conceptual Framework for Collaboration

"Wide strip rolling technology was developed by rival teams in the USA during the mid-1920s. The less successful team at Armco, Ashland, Kentucky was closed to outside influences. Breakthrough came from Columbia Steel at Butler, Pennsylvania which pursued an open pattern of cooperation with equipment suppliers. ... Butler established the dominant design for the next 80 years. The leading equipment supplier at Butler, the United Engineering and Foundry Co., led global sales of the technology for four decades."

Jonathan Aylen (Aylen, 2010)

In the process industries, external actors such as equipment suppliers are important sources of innovation of process technology (Hutcheson *et al.*, 1995; Reichstein and Salter, 2006). Similarly, equipment suppliers are dependent on process firms not only as customers for new process technology, but also for testing and gaining feedback on new prototypes. The incentives for joint development efforts through mutual collaboration are therefore still strong. Referring to the quotation above, a collaborative open mode in innovation is not something new in the process industries, where strong collaborative efforts with equipment manufacturers have always been customary (Aylen, 2010).

In the case of incremental product development, it may not be necessary to involve equipment manufacturers in the early stages of the process innovation process, while in radical product development this may often be the case. In both types of process development it is advisable to have good contacts and collaboration with equipment manufacturers in order to explore new process development opportunities (Lager and Hörte, 2005b). In successful process development, close collaboration with an equipment supplier is often necessary right at the very start of the development of process technology. If the equipment needed is very company-specific (idiosyncratic), it may even be necessary for the process firm to compensate the equipment manufacturer for such development. From the equipment supplier's perspective, the development of new process technology/equipment may be prompted by the identification of customer needs on the world market or internal idea generation and technology push. This chapter introduces a new conceptual model for the life-cycle of process technology/equipment development, relating potential drivers for collaboration and expected outcomes to different phases of the development life cycle (Lager and Frishammar, 2009). Furthermore, a classification matrix for collaboration has been constructed using the dimensions "complexity of equipment" and "newness of equipment" as determinants. The matrix is introduced as part of the conceptual framework to be used in the selection of alternative forms of collaboration (Lager and Frishammar, 2009).

10.1 Equipment supplier/user collaboration — a background and a theoretical platform

When a new production plant is built or an existing one upgraded, it cannot be taken for granted that adequate process technology is available off the supplier's shelves. Rather, it may require a strong commitment on the process firm's part to find competitive production solutions in collaboration with one or more equipment suppliers. The development of such new or improved process technology may be prompted by the process company's need for process development or product development, or both.

Integrating equipment manufacturers into the process firm's innovation processes

As Figure 10.1 shows, external customer demands on the products from the process firm may prompt not only a need for the development of new or improved products but also the development of new process technology to enable the manufacture of such products (Lager, 2008). In addition, customer-driven needs for more efficient and low-cost products may also fuel the development of improved process technology. Such development, however, depends to a large extent on close collaboration with equipment manufacturers.

The development of new or improved process equipment for the process industries is in most cases carried out in close collaboration with one firm or a consortium of process firms, often also targeted as future potential customers for the equipment. In the development stage, the equipment manufacturer is often then the "promoter", most interested in securing development support and collaboration, while in the operation (production) stage the process firm is the one who decides whether and on what premises collaboration should take place. Successful collaborative innovation depends first of all on input of present and future needs for process technology and good ideas in

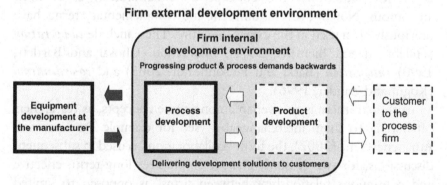

Figure 10.1 The internal and external innovation environment for firms in the process industries. In the external innovation environment not only external customers prevail, but also the suppliers of necessary process technology/equipment (Lager and Frishammar, 2009).

the fuzzy front end (sometimes from different firms). Further on, the execution of an efficient product development phase often uses process firms' production plants for testing or installation of demonstration plants, if such a collaborative approach has been selected.

One way to speed up the product and process development processes for both the process firm and the equipment manufacturer in the future may be to "short-circuit" the product and process innovation chain presented in Figure 10.1. Such a desired effect may be best achieved by stronger integration and improved internal and external cross-functional collaboration, a topic that will be further explored and discussed in the following sections.

A theoretical platform

Collaboration issues have been extensively studied over the past few decades. One side of the literature has focused on collaboration within firms (Frishammar and Hörte, 2005; Kahn, 1996) with a prime focus on collaboration among functions and departments. Other scholars such as Ahuja (2000a, 2000b) have studied external collaboration, with a prime focus on collaboration among firms. While collaborations "within" and "among" firms represent two different ideal types of collaboration situations, the concept of collaboration is in itself ambiguous. Notably, several different and complementary terms have previously been used in the extant literature. These include *co-operation* (Hillebrand and Biemans, 2004), *interaction* (Ghosal and Bartlett, 1990) *integration* (Barki and Pinsonneault, 2005) and *co-ordination* (Kogut and Zander, 1996).

Although there is an overlap among these concepts, as researchers often refer to them interchangeably (see for example De Luca and Atuahene-Gima (2007), the term "collaboration" is used in subsequent discussions, for two reasons. First, it emphasizes long-term, effective and continuous relationships between firms, as opposed to limited transactions and/or exchange of information (Frishammar and Hörte, 2005). Second, the focus is on collaboration between and among firms, rather than within firms. In this context, collaboration is the most commonly used term to characterize joint development efforts.

The literature on intercompany collaboration spans different research domains or traditions. Writings on collaboration between companies have for example been grounded in the resource-based view of the firm (Grant, 1991; Menon and Pfeffer, 2003), the organizational learning literature (Cohen and Levinthal, 1990; Lane *et al.*, 2001), knowledge management (Sveiby, 2001) and product innovation (Chesbrough and Appleyard, 2007). External collaboration may take a variety of forms, ranging from tightly coupled to loosely coupled arrangements. Although an extensive list of forms is presented in the literature, some appear more relevant than others. Specifically, joint ventures, strategic alliances and consortia represent tightly coupled forms, while networks and trade associations (collaborative sectorial research projects) represent more loosely coupled forms (Barringer and Harrison, 2000). A joint venture is created when two or more firms pool a portion of their resources, and create a separate jointly owned organizational unit (Inkpen and Crossan, 1995).

A consortium may be viewed as a special form of joint venture (Brooks *et al.*, 1993), consisting of a group of firms which share similar needs and which then creates a new entity which satisfies this common need (Kanter, 1989). Alliances, on the other hand, represent an arrangement between two or more firms in the form of an exchange relationship that has no joint ownership involved (Dickinson and Weaver, 1997). Networks are constellations organized through social rather than legally binding contracts (Jones *et al.*, 1997). Nevertheless, in collaboration between equipment manufacturers and process firms, the actors can choose from an array of potentially relevant collaboration modes, ranging from tightly coupled to more loosely coupled ones arranged on an informal basis.

Regardless of collaboration mode, however, external collaboration as such has both advantages and disadvantages. Advantages include access to resources, economies of scale, risk and cost sharing, enhanced product development, learning, and flexibility (see for example (Grandori, 1997; Hagedoorn, 1993; Hamel, 1991; Kanter, 1989; Kogut, 1988). Disadvantages typically include loss of proprietary information, increased complexity in management issues, financial risks, increased resource dependence, loss of flexibility and antitrust issues (Doz and Hamel,

1998; Gulati, 1995; Hamel *et al.*, 1989; Jorde and Teece, 1990; Kogut, 1988; Singh and Mitchell, 1996).

Although both the benefits and drawbacks of external collaboration have been discussed, the literature seems biased in the sense that collaboration is usually pictured as being a good thing, while in reality the results of joint collaborative efforts may be both positive and negative, depending on the goals and circumstances of each collaborating partner (Cox and Thompson, 1997; Eriksson, 2008). This is apparent in the process industries, where joint collaboration can lead to major improvements in new process technology, but simultaneously allow unintended knowledge transfer, as when core knowledge is spread to competitors via equipment manufacturers active globally.

As external collaboration contains both positive and negative effects and outcomes, it seems justified to ask why, when and how collaboration should take place, rather than just assuming that firms should collaborate on innovation in new process technology/equipment. Each question will be discussed and elaborated upon in the following.

10.2 Why collaborate — expected outcomes from a collaboration

The external collaborative approach and co-development partnerships in innovation is nowadays often referred to as "open innovation" (Chesbrough and Crowther, 2006), talking about the use of purposive inflows and outflows of knowledge during a distributed development process across organizational boundaries. The motives for defining the business objectives before partnering are stressed and tentatively listed as: increased profitability, shorter time to market, enhanced innovation capability, increased flexibility in R&D and expanded market access (Chesbrough and Schwartz, 2007). It may not be easy to decide to develop new or improved process technology/equipment as a collaborative effort with equipment suppliers and process firms, however, collaborative development may have strong strategic implications for both parties. The driving forces behind such collaboration are not always obvious and may vary, because there are both advantages and disadvantages for each partner.

From the process company's standpoint, collaborative development of new process technology allows the process firm to lower its development risks, assuming the alternative would be to develop in-house, without access to important knowledge provided by an equipment manufacturer. This appears especially important in the situation of a process firm's need for one-off equipment, i.e. when idiosyncratic equipment which does not currently exist on the market must be developed. Next, an early involvement of equipment suppliers may provide opportunities for better adapted or even custom-made equipment that better fits the specific needs of the process firm. In a similar vein, collaborative development provides the process firm an opportunity to become an early user and thus get a first-move advantage over competitors (Liberman and Montgomery, 1988). Finally, new or improved process equipment created through joint collaboration may speed up a process firm's product and process development. Clearly, collaborative development has downsides as well. There is a risk that the firm's core technology may be passed on via equipment manufacturers to competitors (Kytola *et al.*, 2006). As a consequence, proprietary knowledge may diffuse via equipment suppliers to main competitors, who are often customers of the same supplier.

Furthermore, collaborative development projects, unless prompted by specific needs on the part of the process firm, may imply high co-ordination costs and resource utilization, where the latter clearly constitute an opportunity cost. The process firm also runs the risk of production disturbances when installing and testing new equipment which has been jointly created. Finally, close collaboration with an equipment supplier may impose on the process firm a situation where it is "taken hostage", i.e. it constitutes a lock-in effect which may favour the equipment supplier in future purchasing situations (Kanter, 1989).

Equipment manufacturers are also exposed to both advantages and disadvantages when engaging in joint development of new process technology/equipment with a process firm. Advantages to the suppliers are several. Firstly, collaborating with a demanding customer often allows the supplier to improve its development capabilities and its understanding of customer needs (von Hippel, 1986). In a similar vein, access to the customer's ideas and partly tacit knowledge can sometimes be transformed into new or even patentable products. Secondly, joint development projects are often financed by both collaborating parties. Subsequently, the process equipment being developed can be sold to other firms as well, allowing the equipment manufacturer to leverage its New Product Development (NPD) on somebody else's budget (Chesbrough *et al.*, 2006). In addition, the new process technology being developed can typically be more customized with a collaborative arrangement, which increases customer satisfaction but also provides a good reference installation. Also, joint development allows a deeper and more intense relationship through mutual asset specificity.

Last but not least, the opportunity and importance for the equipment manufacturer to develop and test prototypes in a real operating process environment setting is second to none. Disadvantages to suppliers are not to be disregarded. Firstly, development of equipment that is too firm-specific or idiosyncratic may have very limited application areas outside the specific collaborative project, and the equipment firm's alternative use of these allocated resources may be much more profitable in a company perspective. Secondly, failures in joint development and subsequent implementation may hurt the reputation of the equipment manufacturer, which is especially important in the often open information-intensive sectorial communication. Finally, important internal or even proprietary knowledge critical to the equipment manufacture may "leak" via the process firm to other manufacturers of process technology.

10.3 When and how to collaborate: a framework for collaboration

If there is a motive to start collaborative development between an equipment supplier and a process firm, the attendant questions are how such development activities should be set up. A further question is when such commitment during the development project's life time should be distributed to obtain a strong but lean development project.

When to collaborate: picturing collaboration over the development stage of the equipment's life cycle

A project involving a very complex technology and also of a radical newness may span a very long period of time in the process industries. Development cycles over 5–10 years are not uncommon if one includes the necessary time for implementation of the new technology in a new production plant. Figure 10.2 shows a conceptual model of collaboration over an equipment development stage of the life cycle (the production part of the equipment life cycle will not be further discussed here).

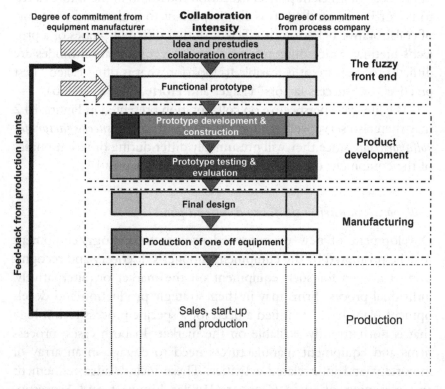

Figure 10.2 A conceptual model of collaboration over the development stage of an equipment life cycle.

The development process in Figure 10.2 has been structured into three distinct phases: the fuzzy front end, product development, and manufacturing of process equipment. Each phase has been further divided into two sub-phases. The process company's commitment and the equipment manufacturer's commitment during the different phases and the collaboration intensities (the inner parts of each rectangle) have been tentatively illustrated by different shadings (the darker, the stronger) in the outer parts of each rectangle. The shaded arrows symbolize the necessary external input for the development work at the fuzzy front end. The black arrow illustrates that necessary input from operating plants is also of importance for the development of new process technology and equipment. How company commitments and their collaboration intensity ought to be in different kinds of projects for efficient project execution today and in the future needs to be further researched. It is also important to understand what sort of collaborative behaviour is efficient during different phases of a project's lifetime. Since outcomes of alternative collaboration modes are difficult to measure, it is feasible to look for what is often called "best practice" or "success factors" (Lager and Hörte, 2005a, 2005b).

The life-cycle perspective on collaboration presented in Figure 10.2 may then also serve well as a framework for studying *success factors for collaboration*, since they will presumably differ during different phases of the equipment development life cycle (see Chapter 14).

Collaboration during the fuzzy front end phase

Development of new or improved process technology/equipment may be prompted by the equipment supplier's discovery and recognition of a need for such equipment on the market or, alternatively, individual process firms may in their strategic production and development plans have identified a need for a specific process technology that is not currently available on the market. In both cases, process firms and equipment manufacturers need to engage in an array of important and interrelated activities. These include idea refinement and screening of ideas (Cooper, 1988a; Elmquist and Segrestin, 2007), early customer involvement (Gassman *et al.*, 2006), senior management involvement (Khurana and Rosenthal, 1998), preliminary technology assessment (Kim and Wilemon, 2002; Verworn, 2006), and assessment of the NPD project vis-à-vis company strategy (Khurana and Rosenthal, 1997).

Development work in the early stages is typically exploratory with many iterative loops. It is, however, important to articulate the needs of the process firm(s) and translate these into a product concept (Cooper, 1988a; Khurana and Rosenthal, 1997). A product definition or functional prototype represents the objective of the development process and is a statement of both technology and customer benefit issues (Montoya-Weiss and O'Driscoll, 2000). Depending on the project's character, this is a phase when preliminary experimental tests take place, complemented in the process industries by modelling and simulation.

Since this phase strongly affects future product performance and cost in the following development phase, it is quite important that the collaborative partners have carefully discussed and agreed upon product specifications and preliminary operating and investment costs for such equipment (Cooper, 1988b), and also how to share the risks.

Collaboration during the product development phase

The different development environments for the development of process technology have been discussed in previous research (Pisano, 1997; Utterback, 1994), in further research about the process innovation work process (Lager, 2000). The iterative loops start in the laboratory (at the equipment supplier's premises or in a process firm's laboratory), going further to pilot plant testing (at the equipment supplier's premises or in the process firm's laboratory) and further to demonstration plant testing. Because of the frequent requirement to test material in larger processed quantities and also to handle the products from the testing, there is often a need for a "process infrastructure" that only a process firm can provide.

Taking a functional prototype into a production environment makes very severe demands on both the equipment supplier and the process firm. The potential operating disturbances of the firm's production processes must be very carefully considered by both parties long in advance, and the necessary risk analysis must have been carried out before testing starts. The privilege for the equipment supplier of operating untested equipment in such production environments must be acknowledged. How long such testing must go on depends, of course, on the character of each project and on the complexity of the process technology, but it typically takes more time than anticipated to develop robust equipment that is not oversensitive to production changes and disturbances.

Collaboration during the manufacturing of process equipment phase

After a successful collaborative product development phase, the commitment for the process firm usually becomes much weaker (see Figure 10.2). However, this is a collaborative phase when there is much important feedback from the process firm to the equipment manufacturer that can improve the final design. This can be in areas such as designing equipment that is easy to operate and with the maintenance costs in focus. For the equipment supplier, this is a phase when the product development work process goes into a progressively more commercial phase and when there are not only strong contacts with the collaborating partner(s) but when marketing of the new equipment becomes more aggressive. We may have here different scenarios, all focusing on the importance of getting a first reference installation to promote further sales:

- The process firm has already purchased the equipment for further installation in a new or already operating plant.
- The process firm may now discuss a possible purchase of such equipment.
- The process firm decides not to purchase the equipment, which puts the equipment manufacturer in a difficult position.

How to collaborate: selecting organizational forms for collaboration

Collaboration between an equipment manufacturer and a process firm may be arranged and decided on a project level, but may also sometimes have to be subordinated to other R&D or strategic considerations. The collaboration between equipment manufacturers and process firms may thus have a hierarchic dimension which is also well worth studying in further research. Holden and Konishi (1996) note that short-term, quick-gain, opportunistic behaviour by firms is unproductive; it will give them the reputation of being bad collaborators and be counterproductive in the long term. Referring to the literature review on collaboration and alternative forms of collaboration, there are today an abundant number of different collaboration forms to choose among, each of them differing in the degree of collaboration intensity (see further Figure 10.2) as well as in legal and other practical consequences. In collaborations between equipment manufacturers and process firms, may some forms be more or less suitable under different circumstances? It therefore seems justified to elaborate upon the criteria for selection of different forms of collaboration, i.e. the key contingencies which determine how collaboration should materialize.

Determinants for different forms of collaboration

There may be a number of possible criteria to consider when selecting a proper form of collaboration during the joint development of process technology/equipment. The time dimension previously touched upon could be one candidate, since some collaborative developments may take a short time but others up to five to ten years. Nevertheless, we argue that "newness" and "complexity" are the two key variables which allow a deeper understanding of when different forms of collaboration are suitable.

Newness of process technology/equipment on the market

In a classification of different kinds of process innovation, the "newness of process innovation on the market" in previous studies (Lager, 2002), and see also Chapter 3. In the categorization of collaboration projects between equipment manufacturers and process firms, "newness of process technology/equipment on the market" was selected as one important determinant, composed of the values low, medium and high.

One way to define a concept is to make an intentional definition, trying to describe what is contained in the concept. Varying degrees of newness and complexity, from low to high, can in this manner be illustrated by examples below from two sectors of the process industries, viz. the petrochemical and mineral industries:

- Low: Well-know process technology/equipment available "off the shelf" through many equipment suppliers (e.g. a valve).
- Medium: Incrementally improved process technology/equipment (e.g. an improved cracker for crude oil).
- High: A radically new process technology/equipment not previously used and possible to protect with patent (e.g. a new natural gas liquefaction plant).

Complexity of equipment/process technology

In considering different contents of the concept "complexity", two alternatives were examined. First of all the "complexity in the development process" itself, which may result in more or less resources needed or different time frames for development, and secondly the "complexity of the product/system" to be developed. The latter alternative was selected as it was easier to grasp and comprehend before development starts. In a buyer-supplier relationship, the complexity of the equipment is one factor that has been recognized as a determinant for collaboration intensity; the greater the complexity, the greater the need for stronger forms of collaboration/cooperation (Eriksson, 2008; Olsen *et al.*, 2005):

- Low: An equipment detail/component (e.g. a wear lining in a grinding mill);
- Medium: A unit process/one operational part (e.g. a new type of mill or classifier);
- High: A complete new process/process plant (e.g. a new grinding system).

A collaboration matrix incorporating complexity and newness is shown in Figure 10.3.

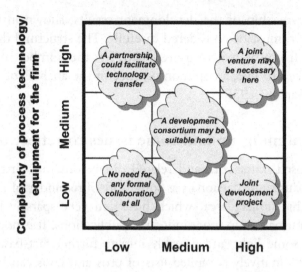

Figure 10.3 The collaboration matrix for the joint development of process technology/equipment. The matrix could first of all be used as a tool in the selection of alternative forms of collaboration. Different forms for collaboration in different parts of the matrix have been proposed. However, which collaborative forms are best suited in different parts of the matrix must be studied in further empirical research.

Ought collaboration on innovation and other collaborative ventures between equipment supplier and process firms to take different forms and be conducted in different ways, all according to both the complexity of the equipment and newness of the equipment? The matrix can thus first of all be used to position collaborative development projects of different kinds to evaluate whether a collaborative approach is of interest at all. Secondly, how strong should such collaboration be (something we call here collaboration intensity)? Looking at the different areas of the matrix, one could speculate that in the lower left corner the needs for formal collaboration are small or even non-existent. On the other hand, going to the upper right corner, there seems to be a need for more tightly coupled arrangements, maybe even a joint venture. In the medium-complex area and medium-to-radical newness areas, a larger development consortium sharing costs and risk may be suitable.

The ownership of the development results, shown in the lower right corner, must be considered carefully. The structural dimensions and scales from the matrix are retained, but the number of areas has been reduced to five, a common practice in the analysis of sociological data (Barton, 1955).

10.4 Summing-up and some issues to reflect upon

The proposed framework and related discussions may first of all be used by industry professionals as some sort of reminder of the importance of this subject area, which has been very sparsely treated in scientific journals or in other industrial publications. It is hoped that it may shed some light and possibly initiate further fact-based discussions. The tentatively compiled lists of pros and cons can be used in internal brainstorming exercises within firms to create more company-specific drivers for collaboration in some sort of ranking order. In collaboration between equipment manufacturers and process firms, such a platform may be jointly discussed and agreed upon in order to ensure long-term open and trustful collaboration.

Reviewing the list of pros and cons from the perspective of the previously presented list of business objectives (Chesbrough and Schwartz, 2007), one can interpret many pros as objectives or expected outcomes of importance which should be identified before a collaborative partnering is established in the innovation stage. Given that there are both pros and cons to close and open collaboration for each party, it seems justified to ask how a win-win situation can be created in such collaborations.

Further on, when such collaboration should take place in different development environments is a question of the highest importance that should be discussed at management level. The conceptual model of the development part of an equipment life cycle is one tool for the collaborating partners for deciding on necessary resource allocations during different stages of a development project's life cycle (degree of commitment), and not only that, but in discussions of how to successfully collaborate in practice during the different phases of the full development life cycle. Choosing among the different organizational forms for collaboration is however something that must be partly guided by company-specific considerations. The proposed matrix can also be used in such discussions, never forgetting the future competitive implications.

Not only the overall time frame, but the intensity of collaboration and commitment of company resources during different phases, will vary between different collaboration projects. The time span for the collaborative development of new or improved process technology/equipment is one dimension that may influence the collaboration intensity and thus also related forms of collaboration.

The conclusions presented by Griffin & Page (1991) support the notion and suggest that the new matrix could be used not only to select suitable forms for collaboration but also to identify related success factors for such collaboration.

- *Do you find that your firm's collaboration behaviour with equipment suppliers can be characterized as "open innovation"?*
- *Has your firm a short-list of preferred equipment supplying partners, in the development of process technology?*
- *In your firm's collaboration with equipment manufacturers, have you considered the different pros and cons for such collaboration, as presented in this chapter?*
- *Collaborative development of process technology is an activity that is often of major importance in process innovation. How much of the process innovation resources are devoted to such collaborations in your firm?*
- *Have you considered how such collaboration should take place and how your firm's resources should be deployed during different phases of the innovation life cycles of such collaboration projects?*
- *Different kinds of collaboration projects are likely to require different forms of collaboration. Try to position your previous, present and future collaboration projects in the matrix presented and discuss the most proper form each collaboration project should take.*

Chapter 11

Process Innovation with the Customers — the Case Study of Application Development at HÖGANÄS

"Application development is primarily the significant development of the customer's use of the supplying company's own products. The development work is an optimization of the customer's production system in the use of the company's products. It may improve the customer's process and/or products. Application development thus implies an involvement in the customer's process and product development."

Proposed definition in Chapter 3

Application development is an important way of building and sustaining long-term customer relations that are supposed to improve customer loyalty and hopefully increase future purchase of company products. Application development can sometimes also maintain margins on products where price creep is common. Improving company market shares thus depends not only on competitive products, but on the collaborative development of the customer's use of those products. In application development the company that makes the products often gradually learns how its products are best utilized by the customer, and the importance of individual product properties. This knowledge must in the future be better internally communicated to product development, and such feedback loops are probably often not so well developed and may need a better formal structure in an application development process (see further Figure 3.2).

Resources for application development are often provided by the company to the customer free of charge, and the customer is the one who is primarily supposed to get the benefits. It is, in very few cases, possible to charge the cost for application development if necessary major investments in specifically dedicated (prototype) production facilities are made. If application development is considered as a service (part of the meta-product) it may be possible in the future to charge the customer for this kind of activity, and regard this as a part of the company's forward integration. This is something that could be of interest for a company to discuss further with its customers. Perhaps the customer would prefer to buy such a service to be in better control of these development activities?

In this chapter application development will first of all be presented together with a theoretical framework including the newly developed application matrix. The framework will afterwards be deployed in a case study of application development at the Höganäs company. The use of such a matrix will be discussed later, and different areas of the matrix will be denominated.

11.1 What is new about a new application — a theoretical framework

Good application development needs a very good knowledge of the customer's production process, while good product development requires an understanding of customer needs, but more importantly of the company's own production process. The individual skills that are important in application development are thus quite different from the skills necessary for good product development. Application development may include minor changes in the company product specifications to facilitate customer use of the product, but not necessitating any kind of company product development. The primary target for application development is usually the improvement of the customer's production process.

The professional application engineer, with intimate knowledge of the company's own product and its use in a customer's production process, is then in an excellent position to set up development activities in collaboration with the customer's production experts. This is primarily in order to improve the customer's production economy, adding value to the customer's production system. An improved use of the company's products on the customer's premises may also give opportunities to improve and add value to the customer's products. This area is not always well explored, since it requires good contacts with the customer's product development teams.

Many authors state that it is largely agreed that world-class R&D performance can no longer be achieved by a company on its own, and that meeting customer requirements increasingly needs R&D collaboration in buyer-supplier relationships (Collins *et al.*, 2002; Dunne, 2008; Hurmelinna *et al.*, 2002). Dow Chemicals says it set up an application centre that focuses on developing solutions with and for customers in the cleaning industry (Seewald and Sim, 2003). Kjeldsteen discusses "the improvement of application opportunities and quality of powder metal products by extended collaboration between powder metal supplier and customer" (Kjeldsteen, 1990). Two recent Australian papers introduce the concept "demand chain" (Rainbird, 2004; Walters, 2006). They propose that the supply chain concept has a corresponding concept and that it will be of equal corporate importance in the future.

A better characterization of different types of development activities is advantageous for the creation of company R&D strategy and selection of a well-balanced portfolio of development projects (Roussel *et al.*, 1991). In the area of innovation, "newness" is by definition a strategic dimension. For product and process innovation the degree of newness is often categorized on a dichotomous scale as "incremental development" and "radical development". Different kinds of innovation can however be better characterized in a corporate perspective by using two-dimensional matrices. The application development can be carried out both with the direct customer in the supply chain but also directly with the "end user". This may also include application development in collaboration with equipment manufacturers to the customer.

In a company's market development efforts, new potential customers for existing products are always sought as well as new potential application areas for the aforementioned products. This is, however, in this book classified as market development.

Selecting the dimensions and scales for the application development matrix

The selected dimensions for the matrix were "newness of customer for the (own) company" and "newness of application for the (own) company". These dimensions were a natural choice in the construction of the application development matrix. The selected scales for the two dimensions were the following:

Newness of customer for the (own) company

> Low: A customer with whom the company has had a previous long-term and trustful collaboration.
> Medium: A fairly new customer for the company.
> High: A completely new customer for the company.

Newness of application for the (own) company (not the customer)

> Low: An application area that is very familiar to the company.
> Medium: Some previous experience in this area for the company.
> High: Completely new, green-field application area for the company.

A new matrix was constructed using the selected dimensions and scales and is presented in Figure 11.1. The percentage of application development within each square of the matrix is a two-dimensional variable. If there is a need for a categorization with a one-dimensional variable, the aggregated percentage of application development in the shaded area can be used as a measure of the "newness of application development" for an R&D organization. One could tentatively use the scale: 0–25 percent very high; 25–50 percent high; 50–75 percent low; 75–100 percent very low. If, for example, the aggregated figure is 80 percent, the newness of application development is very low for the company's total effort in application development.

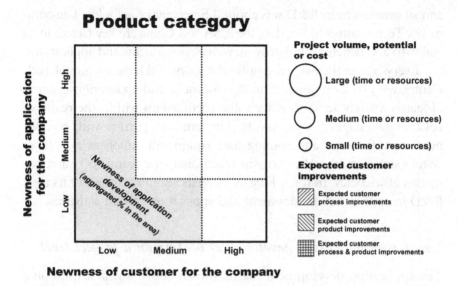

Figure 11.1   The application development matrix; a new classification system. The matrix can be used on a project level, where individual projects can be illustrated with circles of different sizes. The proposed different shading can be used to illustrate expected outcomes of individual application development projects. On an R&D level, the percentage of application development carried out in individual areas gives opportunities for strategic considerations.

11.2 The HÖGANÄS mini-case study

Höganäs is the world's leading supplier of metal powder technology (www.hoganas.com 2008), with an annual turnover in 2007 of about 5800 million Swedish Kronor and with an operating margin of 10.4 percent. Metal powder has been used as an engineering material in the manufacturing of components and consumables for well over 50 years, and the material is now employed in a large number of industry segments. The Höganäs Group is now organized in five regions with full profit-and-loss accountability. Metal powders from Höganäs facilitate designing-in high-value properties into components and consumables whilst ensuring lowest total manufacturing costs. This powers products and systems being delivered to market in a rapid, reliable and cost-effective manner. The overriding aim is to realize the business advantages gained by "designing with powder in mind" from the start. During 2007 the annual investment in R&D was about 2.5 percent of net sales. The company's Tech Centres in Sweden, the USA and China are key factors in its ability to and penetrate and grow new market segments and applications.

Every one of the five geographical regions will have a regional Tech Centre with its own application development and application experts. Höganäs actively seeks to enter value partnerships within the realm of collaborative engineering projects. The aim is to partner with component manufacturers and tooling and equipment suppliers as well as system suppliers, thereby forming multi-discipline teams to realize to-market efficiencies. In total, Höganäs invests about one-third of its total R&D in application development and application-related activities.

Using the application development matrix on a project level

The application development matrix shown above was first used on a project level by selecting one category of products from business area components. The results are presented in Figure 11.2, and a fairly

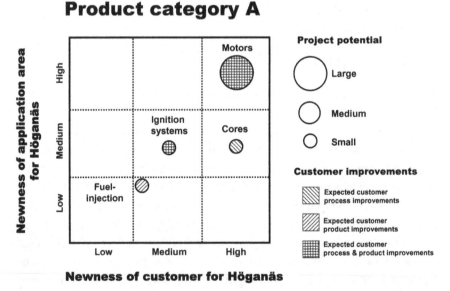

Figure 11.2   Application development projects at Höganäs for one category of products in business area components.

wide spread of application development projects is apparent in the matrix, including a large project in the upper right corner.

Using the application development matrix on a business area level

The application development matrix was primarily aimed at a project-by-project classification and the balancing of different kinds of application development projects. If, however, the aggregated resources spent on application development are distributed in the matrix, it can be used as a more strategic resource allocation tool for application development at a business area level. The distributed application spending for Höganäs' two business areas shown in Figure 11.3

Business area components

The powder metallurgy industry represents some 70 percent of sales. Höganäs supplies high-value powders that are formed into final or

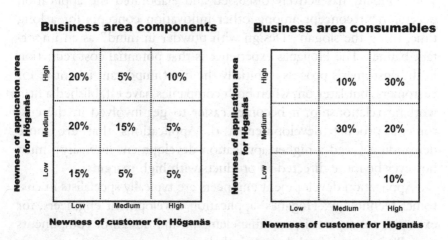

Figure 11.3 Application development in the components and consumables business areas at Höganäs. Aggregated percentage of application development for each business area and for each area of the matrices.

near-net shaped components in the processes of component manufacturers. They then supply components via system or assembly
manufacturers to end users in the automotive industry or for lawn and
garden, hand tool or white goods applications. Total spending on
application development in business area components is aggregated
and distributed in the matrix on the left in Figure 11.3.

Business area consumables

In addition, Höganäs metal powders are used in the welding, brazing,
surface coating, chemical and metallurgical processing industries, primarily by producers of welding, filter and friction consumables, users
of brazing, cutting and coating technologies and producers of food
and feed enrichment. Total spending on application development in
business area consumables is aggregated and distributed in the matrix
on the right in Figure 11.3.

A discussion of application development and case-study results at Höganäs

The company has actively discussed and elaborated the application
development concept among other innovation concepts for a long
time, using the slogan "Design with powder in mind" as an important banner. The Höganäs experience is that potential cost reduction
in the customer's process is initially the most important incentive for
customers, but later on, when both companies have established a firm
working relationship, it becomes easier to get involved in the customer's product development too. Applications that are more
demanding (need a higher application development intensity), must
however be more directed to products with high margins.

Application development engineers are typically specialists in customer technology; Höganäs' application development engineers, for
example, are experts in machine elements and design of components
(e.g. PhD in cogwheel design) and electrical motors.

A matrix was first selected on the project and product category level,
since the results from the matrices on this level can then be further

aggregated into a matrix on a business area level. As such, the Höganäs matrix on a product category level, Figure 11.2, is only one matrix and part of the total matrices on the level of business area components. It must also be observed that the sizes of the circles on the project category level represent project potential (a selection made by Höganäs), but in the business area components' matrix they represent the percentage of allocated resources given to application development. In discussing the matrices and the results, it is however important to remember the dynamics. The application development area named "Motors" in the top right box in Figure 11.2 will gradually decline as time goes by, and finally end up as applications in the bottom left area. The strong potential of this application, and related large necessary resources, must also be considered in the perspective that expected customer benefits are both process and product improvements.

The matrix on the left in Figure 11.3 is the distributed aggregated spending on application development on the level of business area components. In this matrix the "newness of application development" variable adds up to 40 percent, and could be regarded as high newness. The matrix on the right in Figure 11.3 is the distributed aggregated spending on application development on the level of business area consumables. The variable "newness of application development" is here 0 percent, indicating very high newness of application development, which truly reflects the newness of this business area for Höganäs. In five years' time Höganäs will be more bottom-right-heavy, i.e. the balance will have tilted to fewer new applications for Höganäs but still more new customers.

Höganäs' overall experience of using the matrix was that it proved to be a useful tool not only as a reflection of present project activities, but also as a potential tactical tool before starting new application development projects. Trying to work the application development matrix and to delineate possible generic good behaviour and best practice for application development resulted in the "arrow" shown in Figure 11.4. The strategy conclusion is that a fast time to market implies a low-left-to-high-right strategy, as it is easier to first apply low-to-medium changes in both newness to customer and application and subsequently go for the top-right corner. There is also a learning

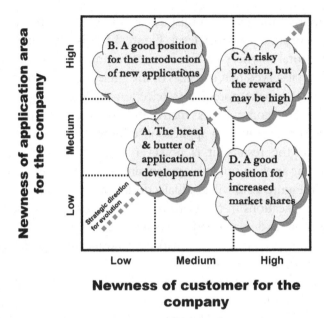

Figure 11.4 A tentative labelling of different matrix areas in the application development matrix.

implication: it is necessary to be able to afford to have sustainable development efforts in order to be successful in the top-right area.

11.3 Balancing application development resources

The structural dimensions and scales from the original application matrix were later on retained, but the number of areas in the matrix has been reduced to four in Figure 11.4, covering most parts of the matrix and taking better account of both dimensions; a common practice in the analysis of sociological data (Barton, 1955). The four areas are:

A. The "bread & butter" of application development. Refining and optimising the production processes for existing customers. Only minor process changes in customer process configurations are necessary. Low investment costs and low risk make this area attractive to the customer. Possible improvements of the customer's products should not be neglected. Although improvements in this area are sometimes difficult to find, the importance of the customers and serving those not-so-new company customers is of the utmost importance. Neglecting existing, profitable customers is a sin.

B. A good position for the introduction of new applications. Introducing a new area of application is often facilitated by a previous good customer relationship. If an existing customer is available for a new application, this is a desirable position for development. Improvements of customer product properties must also be sought.

C. A risky position, but the reward may be high. This is sometimes a difficult area to work in, but may be necessary if the application area does not exist with existing customers (a new-to-the-world application). This is likely to demand plenty of resources in the application development and a high degree of risk. If, however the project is successful, the reward may be high.

D. A good position for increasing market shares. This category of application development is good if increased market shares are sought. The customer selection in strong collaboration with sales and marketing people is essential. Another success factor is the mobilization and use of the company's previous experience in this, rather familiar, application area.

The bottom-left area of the matrix is however not so attractive for application development, as a long relation with an existing customer in the use of a well-known product has probably explored all improvement possibilities. This area may on the other hand then be a more interesting target for customer support. The application development matrix can be used to position portfolios of application development projects or to define and set priorities for a company's application development strategy.

Projects can be marked with circles of different sizes indicating the volumes of project expenditures. Other project characteristics, such as expected type of improvements, type of technology, development time, etc. can also be flagged with different colours.

The project portfolio is well characterized and made more transparent for further analysis, and a more holistic picture of the company's present application development will emerge. At this point it might be appropriate to consider whether the balance between different types of application development activities is good, or whether it should change in the future.

11.4 Summing-up and some issues to reflect upon

A categorization of different kinds of application activities has been made, using a matrix with the dimensions "newness of the application area for the company" and "newness of customer for the company". This theoretical framework was then further introduced and used in a mini-case study at the world's leading metal powder producer Höganäs. The importance of this area of development was confirmed, and the usability of the matrix for a strategic balancing of corporate application development portfolio was acknowledged.

The corporate perspective on application development was discussed, and the final conclusion was that application development is a management issue that has received far too little attention in either corporate management of innovation or academic research.

It is also important to make a clear distinction between market development and application development, as exploration of new customers and new application areas are mainly the responsibility of the market organization (market development), while application development is the collaborative development effort with the customer to improve the customer's use of those products. Generally improved dynamics in corporate behaviour are to be sought in the future: dynamics in product development with application development, in product development with market development, between sales and production, and between central and regional facilities. The overall conclusion from this study is that application development in the process industries is an important area for R&D and because of this should be recognized as a company management issue.

- *If your firm has B2B customers, do you identify application development as an important activity?*
- *How much of your firm's total R&D resources are allocated to application development with the customer?*

- *How much of your firm's total R&D resources are allocated to application development with the end-user?*
- *Using the previously presented application matrix, position your firm's application projects in the matrix and discuss the balance of this project portfolio.*

Chapter 12

Technology Transfer and Startup of New Process Technology

Companies that expected startups to last months instead were still trying to get the mills working smoothly years after the first heat. The more new technologies a mill installed, the longer the startup took. ... Some mills also had the wrong people in place. Despite the million dollars companies spend on the most modern systems, new furnaces, casters, and rolling mills, putting the right people in charge of starting up a new mill is paramount.

Tom Bagsarian (2001)

In discussing process innovation work processes in Chapters 8 and 9, the transfer of process technology to production was recognized as a very important activity. Development of process technology is thus not finished until it is implemented in production. By analogy, just as the final phase of product development is launching the new product on the market, for the process development process the final phase is the implementation and startup of the new production technology. Furthermore, companies' openness for adoption of new inter-company technology is another important aspect of process innovation which will be discussed using the concept of "technology diffusion" (Lager and Frishammar, 2010).

Some experience from startups in the steel and mineral Industry will be presented and can serve as an important reminder of the risks with the introduction of new, untested technology. One should not however conclude that complete risk avoidance is the proper route to follow. When new technology is introduced, preparations before startup can considerably reduce such associated risks.

Startup is all about people interacting with technology, and the organizational aspects of startups sometimes do not get the attention they deserve. Recruitment and training of people in advance, preparing for efficient communication before and during startup, and selecting a proper startup organization are issues that must have top priority when successful startups are desired.

This chapter includes a brief literature review and analysis of publications related to startup of process technology and production plants in the process industries. The results are then discussed particularly from the perspective of safety and implications for industry. Present organizational models and work processes are then critically examined, and a new conceptual model and work process for startup is presented.

12.1 Technology diffusion and transfer — state of the art

The phenomenon of diffusion in physics describes the spontaneous process of dissemination of molecules passing through more or less permeable membranes, and one could view the diffusion of technology in a similar manner as information passing through organizational barriers.

The dissemination of information is at the heart of the concepts of "technology diffusion" and "technology transfer", which are sometimes used interchangeably, causing confusion and hampering communication among practitioners and scholars. In an effort to clarify the distinction between these concepts, Stewart (1987) talks about the company's intentional desire to sell its products or process technology on the market as "diffusion of use". Normally, the company has no desire to sell the embodied technology in the product or the production technology for its manufacture. Such intentional delivery is labelled "technology transfer", when the company, at a suitable price, is prepared to sell the right to produce its new product and the knowledge needed to do so. Diffusion of use is thus restricted by demand, whereas technology transfer is constrained to the supply side (Stewart, 1987).

In this chapter we follow the distinctions proposed by Stewart and consequently use the concept of "technology diffusion" as the spontaneous flow or meandering of information and knowledge about a technology, whereas we use the concept "technology transfer" to label a company's intentional transfer of information and knowledge.

Technology diffusion

The literature on technology diffusion in the area of technology management is vast; in Baptistas's survey of the diffusion of process innovation (1999) he defines technological diffusion as the process by which innovations, whether new products, processes or management methods, spread within and across economies. The "epidemic model" for technology diffusion is based on the spread of information, while the alternative "probit model" for technology diffusion is based on the decision-making processes of individual companies (Baptista, 1999). In sum, three main issues have been studied in the economic literature: the rate of inter-company diffusion, the pace of intra-company diffusion, and early adopters.

Mansfield (1968), one pioneering scholar in this area, studied empirically both the rate at which firms begin to use an innovation (inter-company diffusion) and how rapidly a new technique displaces an old one (intra-company diffusion). The difference between these processes has been further studied in later work (Battisti and Stoneman, 2003) and the conclusion is that inter- and intra-company diffusion of new technology follow different paths over time.

In a study of intra-company diffusion of totally new technology in the pharmaceutical industry (Lessing and Leker, 2006), it was concluded that the diffusion process was a bottom-up process, difficult to promote if no demand from the end-user of the technology existed. Geroski (2000) has listed determinants for diffusion speed as company size, suppliers, technological expectations, learning and search costs, switching costs, and opportunity costs. Companies with newer capital equipment", for example, are less likely to switch to new technology than firms with older equipment, and this is particularly true for capital equipment that is so specialized that the cost incurred in installing it has been sunk. This phenomenon is not unusual in the process industries.

Other factors that måy tend to retard diffusion include the degree to which an innovation is incompatible with existing processes and requires major process changes, the degree to which increased technical skills are required to use the innovation, etc. (Utterback, 1974). Since early adopters have evidently chosen to use the technology despite not having had access to the experience of a previous user, they are somehow different from subsequent users, concludes Geroski (2000). In their study of the diffusion of multiple process technologies, Stoneman and Kwon (1994) conclude that there are considerable cross-technology effects and that the presence of one technology does affect the adoption of another. On the one hand, they conclude, uncertainty connected to the rapid introduction of incremental innovations slows the diffusion process due to expectations of continuing incremental change, while on the other hand increased profitability resulting from early adoption can accelerate the rate of diffusion.

One must, however, also take into account the distinction between the hardware and software aspects of technology and technology diffusion. The hardware is the tool, machine or physical object that embodies the technology, while the software is the information base needed to use it effectively. Although some software can be transmitted impersonally through the user's manual, much of the software of a particular technology is built up from the experience of using it, and at least some of that valuable knowledge is tacit (Rogers, 1983, p. 12). This fact supports the view of collaboration as information-sharing and underlines the importance of good information management in collaborative partnerships.

It is therefore interesting to study the flow of information within companies as a complement to the inter-organizational information-sharing in collaboration between equipment suppliers and process firms. For the process firm, information and knowledge provided by an equipment supplier must be properly disseminated within the firm to different functions and/or sometimes between organizational sub-units or subsidiaries. Since intra-informational dissemination may often be of a more spontaneous nature, it is well covered by the concept of technology diffusion.

From the equipment supplier's point of view, information that is shared from the process firm during the procurement, startup and further operation of the equipment must also be efficiently disseminated within the company (Lager and Frishammar, 2010). Not only must such information and knowledge be transmitted to other sales or commissioning people in related organizational units but, and most important, information must also be conveyed to the internal R&D organization in their further development of improved or new process equipment. It is therefore probably in the future interest of the process firm, too, to manage spontaneous dissemination of information in a more orderly fashion, probably more like well-structured and managed technology transfer.

Technology transfer

If we define technology as the physical object, including the process of making this object and the necessary knowledge to operate the object, the transfer of technology is viewed here as the physical movement of equipment and the transfer of the necessary skills to operate the equipment understanding also embedded cultural skills (Levin, 1993). In two articles, Reisman and Zhao try to establish two different kinds of taxonomies for technology transfer. In the first it is suggested that technology transfer should be classified according to six entities: disciplines, professions, industries, sectors, regions and countries (Reisman, 1989). In the second article and a taxonomy for company-to-company technology transfer, the suggested dimensions to be used for the characterization of different kinds of transactions are duration, cost, nature and modality (the latter is referring to organizational forms for collaboration) (Reisman and Zhao, 1991).

In their study of internal technology transfer and the determinants for success, Leonard-Barton and Sinha (1993) observed that not only the cost, quality and compatibility of the technology mattered. User involvement in the development and the adoption by the developers and users of both the technical system itself and the workplace are important success factors. They observed that a technical system transferred from a development site to a user site always encounters differences in context, equipment, operators' skills, etc., and even if developers successfully meet their original technical objectives, new technology often requires fine-tuning in the operating environment. They also noted that previous experience suggests that users will be more receptive to a new system if they have contributed to its design; more user involvement may not however necessarily be better, since some users may not always want to be involved.

Several other studies of technology transfer note the importance of strong and open personal interactions. In a study of barriers to technology transfer, on a personal unit of analysis, Jung (1980) concludes that an organization that wants to minimize barriers must observe a number of factors, including looking for personalities that facilitate technology and rewarding good technology transfer behaviour.

This is further discussed by Leonard-Barton & Deschamps in their study of managerial influence in implementation of new technology (Leonard-Barton and Deschamps, 1988; Leonard-Barton and Kraus, 1985). A study of technology transfer practice in Japanese corporations (Holden and Konishi, 1996) also emphasizes the importance of human and "soft" factors. Two different kinds of technology transfer relating to the time horizon are identified in this study: "progressive technology transfer" of a more proactive strategic perspective and "defensive technology transfer" more focused on an operational level and to meet short-term needs.

The importance of personal skills and behaviour is well recognized in a study of "spin-In" technology transfer for small R&D biotechnology firms (Galbraith *et al.*, 2004). Initially Malik (2002) concludes that barriers to technology transfer could be overcome by a personnel approach (temporary or permanent transfer of the owner of knowledge to the user group), an observation which is supported by Langrish (1971). It is recognized that barriers or likely-to-inhibit factors are lack of interest in the project, the "not invented here" syndrome, lack of people transfer, no perceived market benefit, lack of trust, no training, lack of incentives, language barriers, and perception of new technology as a threat.

Technology transfer is, however, not a one-way communication process but a reciprocal transfer of knowledge. Malik (2002) presents a model for intra-company technology transfer which he calls the "broadcasting model". Strangely enough, since broadcasting is a one-way form of communication, he points out that this type of technology transfer should be a two-way iterative process and not simply a one-way linear process. Holden and Konishi (1996) also reflect that traditional technology transfer characterized by import of technology is being replaced today by reciprocal collaboration and joint development. They express the need for a more dynamic, inter-active process, balancing internal R&D competencies with those of strategic and virtual partners around the world. A study by Trott *et al.* (1995) recognizes the importance of non-routine activities and effec-tive communication.

In addition, two previous empirical studies have focused on the domain of interest here — process firms. Reporting insights on tech-nology transfer from the Bureau of Mines, Kissell (2000) provides four factors — pressure, path, price and profit — to consider, and notes that transfer of technology is easier than transfer of products. In a short discussion of technology transfer in the petroleum industry, Knight (1984) points out that the fragmented oil business does not allow new equipment to be easily introduced, and joint development cannot occur until a really big company gets involved, highlighting the importance of the size of the equipment manufacturer in the development and introduction of new technology. Apart from noting a general culture of distrusting and rejecting outsiders or new tech-nological ideas, Knight states in his introduction: "The literature of innovation and technology transfer has focused upon successful cases, therefore, almost nothing is written about innovation in the petro-leum industry". In the discussion of mill/supplier relationships in the pulp and paper industry, a number of problems are listed and it is noted that projects that turn out badly for all parties often start with poor contracts. In spite of this it is concluded that a win-win situation certainly could be developed.

12.2 Startup of new or improved process technology

Innovation in the process industries, whether product or process inno-vation, will in its final stage often involve modifications of existing production equipment, new process installations or even the erection of a complete new production plant. The final stage of such activities within the total work process is the implementation and startup. Such a startup of new process technology in a production plant environment can be looked upon as an analogy to product launch on the market in product innovation.

Startup of production plants in the process industries is an important activity which is often discussed in terms of "plant commissioning" (Horsley, 2002). General guidelines for successful such startups are plentiful (Gans, 1976; Gans *et al.*, 1983). In the development of process technology, it is essential to ensure that startup will not be the "weakest link" in the long chain of development activities and cause project failures. As shown in Chapter 7, excellence in startup is a good example of organizational learning. It is not the knowledge of an individual in the firm that counts, but knowledge shared and executed as a joint effort that is the hallmark of a professional and successful startup organization.

Startup — preparing for an extreme event

Taking new plants, production processes, minor unit operations or even a single item of equipment on stream is not only a production and financial risk, but an activity that is always also a safety-critical endeavour (Lager and Beesley, 2010). One should not overlook the installation and startup of even minor equipment integrated in big plants since, regardless of size, there is always a potential of a major process and production disturbance. Consequently, when things do not go according to plan, which they often do during startup, this may strongly influence both the internal and external production environments. Good planning before a startup is thus extremely important and a success factor of the highest rank (Callow, 1991; Meier, 1982; Leitch, 2004b). In the execution of new ventures such as startup of new process technology or plants, it is not the plan itself that is of major importance but the planning.

In Table 12.1 a compilation of different kinds of risks during startup is shown, which can serve as an embryo for internal risk analysis. Some

Table 12.1 A Typology of Risk Areas in Startup of New or Improved Process Technology in the Process Industries. Areas Marked in Italics Require Immediate Action or Temporary Shutdown.

Environmental risks	*External environment — emissions beyond limits*
	Internal environment — not healthy to operate
Technology risks	Wrong selection of process technology (too risky)
	process technology not functioning well
Financial risks	Design & construction too costly
	Operation not profitable
	Delayed start-up
Operational risks	No design throughput
	Product quality not acceptable
Personal risks	*Not safe to operate*
	Climate during start-up too stressful

of those risks have however already been taken and decided upon in the pre-study of the project, while others may be acted upon in the preparations for the startup. Unfortunately it is often during startups that the most difficult things occur which not have been identified in the previous risk analysis.

The importance of advanced risk analysis in plant commissioning is stressed by Cagno *et al.* (2002). It is further advocated by Leitch (2004a) that effective new plant startup increases the asset's net present value, and an integrated work process and upfront planning for the startup is the recommended solution.

Startup performance

It is important that stakeholders and the project management team agree on performance criteria for a process technology or production plant early on in the pre-study, design and procurement phases. The success of an investment project and the associated startup should be compared to such preselected plant performance criteria. Sometimes the outcomes of investments in new process technology and process plants are unfortunately often measured by a few simplistic measures.

Some additional measurables for design and startup excellence are presented below:

- Investment costs (profitability, keeping project budget, etc.)
- Total operating costs (maintenance cost for the plant, raw material yields, etc.)
- Product quality (product performance, product uniformity, etc.)
- Production flexibility of the plant.
- Production capacity (throughput etc.)
- Availability.
- Ease of future upgrading of the production system.
- Startup according to plan.
- Lead time until production targets are reached.

It is essential that the startup management team has developed and secured a realistic and well-founded startup plan accepted by the stakeholders (Figure 12.1). Different industry sectors and individual firms may however have different views on how fast design targets should be met.

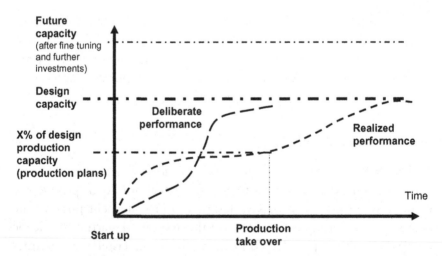

Figure 12.1 Startup curves. It is important to agree well in advance when production takeover should occur.

Every startup has a startup curve in the time dimension. Referring to and borrowing from Mintzberg's strategy concepts in Chapter 6, we can have a deliberate start-up curve, an emergent startup curve and a realized startup curve, see further Figure 12.1. The deliberate startup curve is the one the startup management will try to follow (the emergent startup curve is then the unexpected one). Experience tells us that startup procedures that are well and carefully executed and systematically carried out are likely to be more efficient in the long run that forced startups of a crash project character.

Two papers by Agarwal and colleagues present the experiences from startup of a large number of mines and processing plants (Agarwal *et al.*, 1984; Agarwal and Katrat, 1979). In the conclusions it is stated that:

"The projections of cash flow from new ventures must be based upon realistic start-up rates and allow for unexpected delays in achieving design capacity. The degree of success for a new mine or plant start-up can be dramatically improved by investing sufficient resources in preparing for operation. These resources are small by project standards and, on average, represent only about 1 percent of total project cost."

The time for a plant to get to full production is sometimes also referred to in the process industries as the ramp-up period (Nendick, 2002). The importance of performance testing and how to organize this kind of activity depends on people and planning. The startups of large-scale steelmaking projects in the years 1995–2000 were largely dismal (Bagsarian, 2001). Figure 12.2 shows that none of the plants had reached design capacity within the first year and only one after two years. Bagsarian (2001) shows that the causes for the related slow startups were mainly investment in new, untried process technology:

"Gallatin hired contractors to build the plant and then hired its own people to run it. This was one of the biggest mistakes they made. ...
It's important to have the guys that are going to run the equipment involved from the very beginning, with the design and construction and obviously in start up."

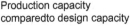

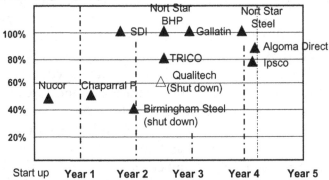

Figure 12.2 Startup of 11 steel plants between 1995 and 2000 in the USA (Bagsarian, 2001).

In an analysis of a large number of startups in the mineral industry McNulty (1998) came to the following conclusion:

"There are many potential pitfalls in the commercialisation of innovative technology. The failures often are remembered longer than the successes, thereby serving as deterrents to subsequent attempts. The Mineral Industry's future depends on effective innovation. So it is essential that the causes of failure are understood."

Figure 12.3 presents the rather depressing history of the startup of 41 base-metal production plants.

Characteristics of Group A:

- Reliance on mature technology.
- Equipment similar in size and duty.
- Thorough pilot-scale testing of potentially risky unit operations.
- If technology was licensed, there were many prior licences.

Characteristics of Group D:

- If any pilot-scale testing was conducted it was not for confirming process parameters.

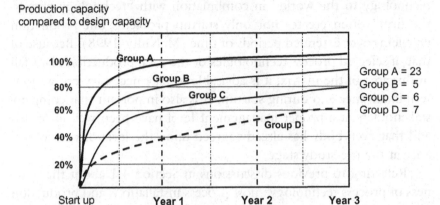

Figure 12.3   Startup of 41 base-metal plants (1965–1995). The production units were flotation plants and smelters (McNulty, 1998).

- Equipment was downsized in response to projected cost overruns.
- Process flowsheets were unusually complex with prototype equipment in two or more critical unit operations.
- Process chemistry was misunderstood.

Three of the plants in this group were closed down within 36 months of startup beginning. The problems presented in the startup of the base-metal plants may give the impression that few startups in the process industries are successful or on schedule. The modernization of the Lafarge Richmond cement plant proves the contrary (Melick, 2000).

Selection of process technology before startup

Startup success depends to a large extent on the selection of proper process technology and professional design of the production plant. Even the most professional startup organization will fail if the selected process technology is not fully developed or poorly applied. Experience thus shows that technology selection is of the utmost importance and may be the single most important factor influencing startup success (Bagsarian, 2000, 2001).

The top-right corner of the process matrix in Figure 4.2 presented in Chapter 4 is labelled "radical & risky". It is well proven that "new technology to the world" in combination with "technology new to the firm" often creates not only startup problems but production problems over extended periods of time (McNulty, 1998). Because of that, if selected process technologies or the total production plant fall in this area of the matrix, it is advisable to take necessary precautions well in advance and during startup. It is also important that company stakeholders at a proper management level have been well informed and that everybody has already agreed upon the level of risk acceptance at the pre-study stage.

Referring to previous disscussions in Section 4.1 about the newness of process technology, most process installations and production plants have a spectrum of different technologies, each one more or less well tried. In the process industries the total production system is thus often what one could characterize as a "one-off". Such systems are not really well known until they have been tested in operation. But since the selection of process technology and associated risks are already decided upon in the pre-study, it difficult to back out in the later phases when problems occur during startup.

In investments in new production plants one does not however want to get stuck with production technologies that are old, outdated and uncompetitive! On the other hand the associated risks of new and untried process technology should not be underestimated. The balance is delicate, and certainly an issue for more top management considerations, since calculations are often based on the assumption of a smoothly operating production plant, short startup period, and no need for supplementary new investments.

Procurement and collaboration with equipment suppliers

There are at present a number of different collaborative modes and procurement philosophies to choose among when investing in new production processes, from in-house design and procurement of individual equipment, through procurement of specified process functionalities and product properties, to turnkey plant procurement.

In the successful startup of the Rainy River's Plant at Kenora in Canada, a multi-disciplinary team designed an efficient deinking pulp mill with quick startup (95 percent of design capacity by the fifth month after startup), high quality pulp and minimal effect on the mill's existing newsprint machines (Frazier *et al.*, 1996). The following purchasing philosophy was selected:

> "An important early decision made by the team was to purchase the best equipment for each application instead on relying on a single-source supplier. The inclusion of performance guarantees for each piece of equipment, coupled with the experience of team members, negated the requirements for system performance guarantee. A surprising benefit of this "cherry picking" approach was a significant savings in equipment costs."

Turnkey is not always the one and only best solution; the following is a presentation from the IPSCO company in its lawsuit against Mannesmann (Bagsarian, 2001):

> "Not only was completion of the project delayed for an extraordinary and wholly unanticipated amount of time, but neither the facility components, nor the plant in general, has the quality, fitness for purpose, productivity, and performance as represented, warranted and guaranteed."

Different forms of procurement may suit different kinds of technologies and company environments, and the conclusion is often that the more familiar the firm is with a technology, the more procurement of individual items from selected suppliers is preferable. On the other hand, when a firm is not so familiar with a selected technology, turnkey procurement may be more advantageous.

Training before startup

Training before startup is of course extremely important, and it is rather surprising that this is sometimes neglected. The importance of training in complex startups is advocated by Coan and Rankin (1997), and they claim that investment in workforce development was the road to success in startup at Appleton Papers. Customizing a proper training programme for different kinds of personnel is recommended (Bossard and Rometti, 1982; Seal and Turner, 1982), not only for plant operators but also maintenance crews, etc. Such training can be given for example by:

- The company's own professional instructors.
- External professional instructors.
- Experienced plant operators.
- Project members.
- Equipment suppliers.

The selected form such training should have is another matter and one can choose between several options: classroom; dynamic simulation (Rutherford and Persard, 2003); training in pilot plants; training on site during erection; training on site during commissioning, etc.

Information and communication before and during startup

The issue of communication and information sharing before and during startup must be carefully planned well in advance. Issues such as how meetings should be scheduled during startup, how records and minutes are to be kept, the possible use of a web-based information system and large screens in the control-room and in the plant for communication should be considered.

When reviewing publications in the area of startup it is discovered that startup of new or improved process technology or production plants often focuses perhaps too much on the design, construction and commissioning of the equipment — an engineering perspective. Successful startups are however also, and to a high degree, activities that ought to have more of an organizational perspective, because those issues are no less important. Focusing on "putting equipment on stream" obscures the final outcome. The process technology and the process plant are only the means for production of high-quality products in a cost-efficient production system! It is advisable to have both the perspective of startup of equipment and the complementary perspective of process and product startup in mind when planning and evaluating new investments in production technology.

12.3 A startup organization — a new conceptual framework

The importance of building a resourceful startup organization well prepared to handle the extreme environment of such an event is second to none. The startup organization is then a total mobilization of all necessary and available resources within and outside the company. It is not difficult to demobilize those resources if the startup runs very smoothly, but on the other hand it is very difficult to generate more resources during startup if and when problems occur. A study of the transfer of new biotechnological processes from research and development to manufacturing highlights the importance of management of a closely integrated technology transfer team with membership from development, manufacturing, engineering, quality and validation (Gerson and Himers, 1998).

The different categories of people one may consider including in the start up organization are for example:

- Plant operators (reinforced shifts)
- Maintenance (reinforced and also on shift?)
- Plant engineers and process engineers (reinforced and on shift?)
- Equipment suppliers (on call)
- Construction crew (on call)
- External consultants (on shift)
- Start up leaders (on shift and with backup)
- Other personnel as required.

It is often usual and recommended for the startup leader to have a "flying squad" of very experienced personnel available for use when major problems are encountered during startup. The structural perspective on the startup organization will be discussed more in depth in the following. In such an organization the importance of selecting the startup leader (see the quotation at the beginning of this chapter), is paramount.

The importance of selecting the leaders is stressed in a discussion of the roles of the process development group and manufacturing in biopharmaceutical process startup (Goochee, 2002). Here it is recommended that process development and plant startup leaders (two startup leaders on 12-hour shifts) are selected nine months prior to startup. The personal qualifications and the leader's personality will also to a large extent influence the climate during startup. It must also be crystal clear what responsibilities the leader(s) should have during startup, to whom they should report and their availability during startup. Because of the need for quick decisions and action during this period, shift-working startup leaders are sometimes preferred. If the project manager of pre-studies, design and plant erection can later assume responsibility for being the startup leader and afterwards become the plant superintendent, that may be the most desirable organizational solution.

Alternative startup organizational solutions and best practice from the past

There are various possible ways to organize startups. One is to rely fully on the project organization and let the project manager take the role of startup leader. The project organization will then be in charge of plant operation during commissioning and subsequent operation. Plant operators recruited for further plant operation are then "borrowed" by the project organization. When the plant is operating smoothly, it is handed over to production. Another alternative is to rely fully on the line organization, trusting that the project is well integrated in the production organization. The production organization is then in charge during startup, and the project organization assists the line organization. Most common though is a handover from a project organization to the line organization when the pre-commissioning is over and when it is time to "press the start button" and putting the plant on-stream on a shift basis. There are naturally a number of pros and cons for all three alternatives. It is not often that the line organization has the resources to supervise a large investment project, and there is consequently a high risk of project mismanagement.

The solution of letting the project organization remain in charge during startup is often complicated because of union or other organizational problems with the ownership of equipment. Handover sometimes works but is often a source of startup problems. Line organizations are then not involved in the design and commissioning, while the project organization has a tendency to disappear too soon during startup. None of the presented solutions seems to work well, and as a consequence the following alternative solution is proposed.

A *new conceptual framework for a startup organization*

It is often found that the organization of startups is not given enough attention in connection with investments in new process technology and in new production plants. The need for a separate project management organization before startup is often well recognized, and the consecutive takeover by the production organization is only natural. But how to handle the "fuzzy in between"? Problems often occur in handovers and in organizational interfaces, and this interface is no exception.

Sometimes one imagines that management is hoping that startup is just a matter of "pressing the button", after which everything will run smoothly from the start, so there is no need for any special arrangements. As we have seen from the publications presented in this chapter, nothing could be more wrong. Startup is an extreme event and demands an extreme organizational solution. It is thus proposed that the two structures "project organization" and "production organization" should be supplemented by a very distinct and formal intermediate "startup organization" as shown in Figure 12.4.

From the start of commissioning activities, and naturally as a gradual introduction beforehand, the startup organization should take full responsibility for all activities (Lager and Beesley, 2010). In a "merger" of the project organization and future production organization the startup leader will then be fully in charge of this exceptionally strong and well-integrated organization. It must be reinforced according to suggestions in the previous presentation, and there should be no question of who is in control. The full team is

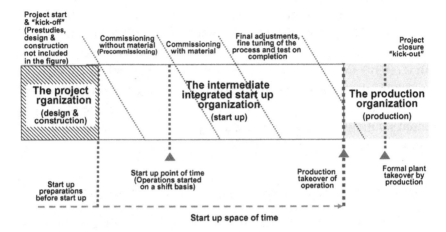

Figure 12.4   A new conceptual model for organization of the startup of new or improved process technology (Lager and Beesley, 2010).

gradually mobilized during pre-commissioning, and in full force during commissioning. This startup organization then stays in control until the plant is operating smoothly. It may take a few weeks or a few months. When agreed upon in advance (performance criteria), the production organization takes over operation of the plant. After the plant has been in operation for some time and the list of outstanding construction items has been seen to, the production organization can finally and formally take over the production plant from the project.

The empirical results from the explanatory-survey to experienced startup leaders in different sectors of the process industries confirmed the usability of such a model. Comments from respondents were:

"This looks like a good model for startup of larger installations and especially if this is introducing new technology."

"This model may be somewhat more expensive but the additional cost will probably be a good investment in the long run."

12.4 Summing-up and some issues to reflect upon

This chapter starts with a review of the concepts of technology diffusion and technology transfer. Those activities have been discussed in previous chapters on work processes for process innovation, and it was then recognized that they are of the utmost importance in well-functioning work processes. In the following Part 5 of this book you will also discover that this area was also given top ranking in an evaluation of success factors for process innovation by industry professionals. Startup of new process technology and production plants is thus largely a matter of technology transfer: transfer of technology and knowledge from equipment manufacturers to production, and transfer of new process concepts from R&D to production.

The message in this chapter is that process innovation and introduction of new, unfamiliar process technology to the firm is and always will be associated with a certain amount of risk. Total risk avoidance is however not going to give the firm competitive and possibly new core process technology which already was highlighted with the citation from Hans Rausing in the introduction. The solution to the problem is instead to prepare well in the introduction of new technology and let an efficient startup organization ensure that the startup will be successful. When new technology is introduced, it is thus important to see that it is properly tested in advance in pilot plants or demonstration plants, that design solutions are professional, and that alternative solutions for handling startup problems are discussed at an early stage. Finally, the organizational aspects should be in focus in the planning of the startup, and the startup organization will be the key to getting the selected technology on stream in an efficient manner.

- *With reference to the technology diffusion concept, how "open" is your firm for influx of new technologies from sources outside the firm?*
- *If the diffusion of technology goes in the opposite direction, how careful is your firm to protect core technology and avoid unintentional leakage of technology to competitors?*
- *How good is your firm's R&D organization at transferring developed technology to production?*
- *Reflect upon the presentation of technology transfer, and consider how your firm could improve technology transfer performance!*

- *How successful is your firm at starting up new process technology or new production plants?*
- *Does your firm try to accumulate experience from previous startups and from seasoned startup leaders in an organizational learning approach?*
- *Reflect upon the previously presented aspects of startup (Section 12.2) and rank them in order of importance, and further benchmark your firm's ability in those areas!*
- *The selection of process technology has been recognized as one of the most important aspects of startup success. What is your firm's policy in technology selection?*
- *Is it possible to select process technology with no risk and still be competitive and cost-efficient in the future?*
- *Consider the organizational solutions for startup presented here, and try to make a list of pros and cons for the different alternatives!*

References

Agarwal, JC, Brown, SR and Katrak, SE (1984). Taking the sting out of project startup problems. *Engineering and Mining Journal*, 62–76.

Agarwal, JC and Katrat, FE (1979). *Economic Impact of Startup Experiences of Smelters.*

Ahuja, G (2000a). Collaboration networks, structural holes, and innovation: a longitudinal study. *Administrative Science Quarterly*, 45, 425–455.

Ahuja, G (2000b).The duality of collaboration: inducements and opportunities in the formation of inter-firm linkages. *Strategic Management Journal*, 21, 317–343.

Auranen, I (2006). METSO minerals: Needs for R&D and education; strength through co-operation: the yearly promote workshop; Innovation in the process industries. Luleå.

Aylen, J (2010). Open versus closed innovation: development of the wide strip mill for steel in the USA during the 1920s. *R&D Management*, 40(1), 67–80.

Bagsarian, T (2000). The mood brightens at Gallatin. *Iron Age New Steel*, 16, 8–27.

Bagsarian, T (2001). Avoiding startup stumbles. *Iron Age New Steel*, 17, 16–19.

Baptista, R (1999). The diffusion of process innovations: a selective review. *International Journal of the Economics of Business*, 6, 107–129.

Barki, H and Pinsonneault, A (2005). A model of organizational integration, implementation effort, and performance. *Organization Science*, 16, 165–179.

Barringer, B and Harrison, J (2000). Walking the tightrope: creating value through interorganizational relationship. *Journal of Management* 26, 367–403.

Barton, AH (1955). The concept of property-space in social research. In *The Language of Social Research*, PF Lazarsfeld and M Rosenberg (eds.), The Free Press.

Battisti, G and Stoneman, P (2003). Inter- and intra-firm effects in the diffusion of new process technology. *Research Policy*, 32, 1641–1655.

Bossard, CA and Rometti, RT (1982). Design and installation of comminution circuits. In *Operator training and startup for semiautogenous grinding circuits*. Littleton: Society of Mining Engineers.

Brooks, M, Blunden, R and Bidgood, C (1993). Strategic alliances in the global container transport industry. In *Multinational Strategic Alliances*, R Culpan (ed.), New York: IB Press.

Cagno, E, Caron, F and Mancini, M (2002). Risk analysis in plant commissioning; The multilevel hazop. *Reliability Engineering and System Safety*, 77, 309–323.

Callow, MI (1991). Start-up autogenous grinding circuits successfully. *Chemical Engineering Progress*, 87(5), 45–50.

Chesbrough, H and Appleyard, MM (2007). Open innovation and strategy. *California Management Review*, 50, 57–76.

Chesbrough, H and Crowther, AK (2006). Beyond high tech: early adopters of open innovation in other industries. *R&D Management*, 36, 229–236.

Chesbrough, H and Schwartz, K (2007). Innovating business models with co-development partnerships. *Research Technology Management*, January–February, 55–59.

Chesbrough, H, Vanhaverbeke, W and West, J (2006). *Open Innovation: Researching a New Paradigm*. Oxford: Oxford University Press.

Coan, W and Rankin, A (1997). Training paves the way coating complex startup. *Papermaker*, 79, 42–45.

Cohen, WM and Levinthal, DA (1990). Absorptive capacity: a new perspective on learning and innovation. *Administrative Science Quarterly*, 35, 128–152.

Collins, R, Dunne, T and Michael, OK (2002). The "locus of value": a hallmark of chains that learn. *Supply Chain Management: An International Journal*, 7, 318–321.

Cooper, RG (1988a). Predevelopment activities determine new product success. *Industrial Marketing Management*, 17, 237–247.

Cooper, RG (1988b). *Winning at New Products*. London: Kogan Page.

Cox, A and Thompson, I (1997). Fit for purpose contractual relations: determining a theoretical framework for construction projects. *European Journal of Purchasing & Supply Management*, 3, 127–135.

De Luca, L and Atuahene-Gima, K (2007). Market knowledge dimensions and cross-functional collaboration: examining the different routes to product innovation performance. *Journal of Marketing*, 71, 95–112.

Dickinson, P and Weaver, K (1997). Environmental determinants and individual-level moderators of alliance use. *Academy of Management Journal*, 40, 404–425.

Doz, Y and Hamel, G (1998). *Alliance Advantage*. Boston: HBS Press.

Dunne, AJ (2008). The impact of an organization's collaborative capacity on its ability to engage its supply chain partners. *British Food Journal*, 110, 361–375.

Elmquist, M and Segrestin, B (2007). Towards a new logic for front end management: from drug discovery to drug design in pharmaceutical R&D. *Creativity and Innovation Management*, 16, 106–120.

Eriksson, PE (2008). Achieving suitable coopetition in buyer-supplier relationships: The case of AstraZeneca. *Journal of Business-to-Business Marketing*, 15, 425–454.

Frazier, WC, Scott, S and Beach, T (1996). Project team starts up Rainy River's recycled pulp plant at Kenora, Ont. *Pulp & Paper*, 70, 69–73.

Frishammar, J and Hörte, SÅ (2005). Managing external information in manufacturing firms: the impact on innovation performance. *Journal of Product Innovation Management*, 22, 251–266.

Galbraith, CS, Denoble, AF and Ehrlich, SB (2004). "Spin-in" technology transfer for small R&D bio-technology firms: the case of bio-defense. *Journal of Technology Transfer*, 28, 377–382.

Gans, M (1976). The A to Z plant startup. *Chemical Engineering*, March, 72–82.

Gans, M, Kiorpes, SA and Fitzgerald, FA (1983). Plant startup-step by step. *Chemical Engineering*, 90(20), 74–100.

Gassman, O, Sandmeier, P and Wecht, CH (2006). Extreme customer innovation in the front-end: learning from a new software paradigm. *International Journal of Technology Management*, 33, 44–66.

Geroski, PA (2000). Models of technology diffusion. *Research Policy*, 29, 603–625.

Gerson, DF and Himers, V (1998). Transfer of processes from development to manufacturing. *Drug Information Journal*, 32, 19–26.

Ghosal, S and Bartlett, CA (1990). The Multinational Corporation as an Interorganizational Network. *Academy of Management Review*, 15, 603–625.

Goochee, CF (2002). The roles of a process development group in biopharmaceutical process startup. *Cytotechnology*, 38(1–3), 63–76.

Grandori, A (1997). An organizational assessment of interfirm coordination modes. *Organizational Studies*, 18, 897–927.

Grant, RM (1991). The resource-based theory of competitive advantage: implications for strategy formulation. *California Management Review*, 33, 114–135.

Griffin, A and Page, AL (1991). PDMA Success measurement project: Recommended measures for product development success and failure. *Journal of Product Innovation Management*, 13, 478–496.

Gulati, R. (1995). Does familiarity breed trust? The implications of reported ties on contractual choices in alliances. *Academy of Management Journal*, 38, 85–111.

Hagedoorn, J (1993). Understanding the rationale of strategic partnering: Interorganizational modes of cooperation and sectoral differences. *Strategic Management Journal*, 14, 371–385.

Hamel, G (1991). Composition for competence and inter-partner learning within international strategic alliances. *Strategic Management Journal*, 12, 83–103.

Hamel, G, Doz, Y and Prahalad, C (1989). Collaborate with your competitors and win. *Harvard Business Review*, 89, 133–139.

Hillebrand, B and Biemans, WG (2004). Links between internal and external cooperation in product development: An exploratory study. *Journal of Product Innovation Management*, 21, 110–122.

Holden, PD and Konishi, F (1996). Technology transfer practice in Japanese corporations: meeting new service requirements. *Technology Transfer*, Spring–Summer, 43–53.

Horsley, D. (2002). *Process Plant Commissioning*. Rugby, Warwickshire: Institution of Chemical Engineering.

Hurmelinna, P, Peltola, S, Tuimala, J and Virolainen, V-M (2002). Attaining world-class R&D by benchmarking buyer-supplier relationships. *International Journal Production Economics*, 80(1), 39–47.

Hutcheson, P, Pearson, AW and Ball, DF (1995). Innovation in process plant: a case study of ethylene. *Journal of Product Innovation Management*, 12, 415–430.

Inkpen, A and Crossan, M (1995). Believing is seeing: Joint ventures and organizational learning. *Journal of Management Studies*, 32, 595–618.

Jones, C, Hesterly, W and Borgattl, S (1997). A general theory of network governance: exchange conditions and social mechanisms. *Academy of Management Journal*, 22, 911–945.

Jorde, T and Teece, D (1990). Innovation and cooperation: Implication for competition and antitrust. *Journal of Economic Perspective*, 4, 75–96.

Jung, W (1980). Barriers to technology transfer and their elimination. *Journal of Technology Transfer*, 4, 15–25.

Kahn, KB (1996). Interdepartmental integration: A definition with implications for product development performance. *Journal of Product Innovation Management*, 13, 137–151.

Kanter, RM (1989). *When Giants Learn to Dance*. New York: Simon & Schuster.

Khurana, A and Rosenthal, SR (1997). Integrating the fuzzy front end of new product development. *Sloan Management Review*, 38, 103–120.

Khurana, A and Rosenthal, SR (1998). Towards holistic "front ends" in new product development. *Journal of Product Innovation Management*, 15, 57–74.

Kim, J and Wilemon, D (2002). Focusing the fuzzy front-end in new product development. *R&D Management*, 32, 269–279.

Kissell, FN (2000). Insights on technology transfer from the Bureau of Mines. *Journal of Technology Transfer*, 25, 5–8.

Kjeldsteen, P (1990). Improving the application opportunities and quality of P/M products by extended collaboration between P/M supplier and customer. *Advances in Powder Metallurgy*, 131–140.

Knight, KE (1984). Technology transfer in the petroleum industry. *Journal of Technology Transfer*, 8, 27–34.

Kogut, B (1988). Joint ventures: theoretical and empirical perspectives. *Strategic Management Journal*, 9, 310–332.

Kogut, B and Zander, U. (1996). What do firms do? Coordination, identity and learning. *Organization Science*, 7, 502–518.

Kytola, O, Hurmelinna-Laukkanen, P and Pynnonen, M (2006). Collision or co-operation course-pulp and paper industry vs. information & communication technology. In *The R&D Management Conference: Challenges and Opportunities in R&D Management — Future Directions for Research*, J Butler (ed.), Newby Bridge, UK.

Lager, T (2000). A new conceptual model for the development of process technology in process industry. *International Journal of Innovation Management*, 4, 319–346.

Lager, T (2002). A structural analysis of process development in process industry–A new classification system for strategic project selection and portfolio balancing. *R&D Management*, 32, 87–95.

Lager, T (2008). Using multiple progression QFD for roadmapping product and process related R&D in the process industries. *14th International Symposium on Quality Function Deployment*, Beijing: China.

Lager, T and Frishammar, J (2009). Collaborative development of new process technology/equipment in the process industries: In search of enhanced innovation performance. In *R&D Management conference*, J Butler (ed.), Vienna.

Lager, T and Hörte, S-Å (2005a). Success factors for the development of process technology in process industry Part 1: A classification system for success factors and rating of success factors on a tactical level. *Int. J. Process Management and Benchmarking*, 1, 82–103.

Lager, T and Hörte, S-Å (2005b). Success factors for the development of process technology in process industry Part 2: A ranking of success factors on an operational level and a dynamic model for company implementation. *Int. J. Process Management and Benchmarking*, 1, 104–126.

Lane, P, Salk, JE and Lyles, MA (2001). Absorptive capacity, learning and performance in international joint ventures. *Strategic Management Journal*, 22, 1139–1161.

Langrish, J (1971). Technology transfer: Some British data. *R&D Management*, 1, 133–135.

Leitch, J (2004a). Effective new plant startup increases asset's net present value. *Hydrocarbon Processing*, 95–98.

Leitch, J (2004b). Successful LNG terminal starts with detailed plan. *Pipeline and Gas Journal*, 231, 58–61.

Leonard-Barton, D and Deschamps, I (1988). Managerial influence in the implementation of new technology. *Management Science*, 34, 1252–1265.

Leonard-Barton, D and Kraus, WA (1985). Implementing new technology. *Harvard Business Review*, November–December, 102–110.

Leonard-Barton, D and Sinha, DK (1993). Developer-user interaction and user satisfaction in internal technology transfer. *Academy of Management Journal*, 36, 1125–1139.

Lessing, M and Leker, J (2006). The intra-firm diffusion of synergetic R&D technologies in the chemical and pharmaceutical industry. In *R&D Management conference*, J Butler (ed.), Newby Bridge, UK.

Levin, M (1993). Technology transfer as a learning and development process: an analysis of Norwegian programmes on technology transfer. *Technovation*, 13, 497–518.

Liberman, M and Montgomery, D (1988). First-mover advantages. *Strategic Management Journal*, 9, 41–58.

Malik, K (2002). Aiding the technology manager: a conceptual model for intra-firm technology transfer. *Technovation*, 22, 427–436.

Mansfield, E (1968). *Industrial Research and Technological Innovation — An Econometric Analysis*. New York: WW Norton & Co Inc.

McNulty, TP (1998). Develop innovative technology. *Mining Engineering*, 50, 50–55.

Meier, FA (1982). Is your control system ready to start up? *Chemical Engineering*, February, 76–87.

Melick, JH (2000). Modernization of the Richmond Cement Plant.

Menon, T and Pfeffer, J (2003). Valuing internal versus external knowledge: explaining the preference for outsiders. *Management Science*, 49, 497–513.

Montoya-Weiss, MM and O'Driscoll, TM (2000). From experience: applying performance support technology in the fuzzy front end. *The Journal of Product Innovation Management*, 17, 143–161.

Nendick, RM (2002). Plant ramp up and performance testing. *Mineral Processing Plant Design*, 2, 2285–2289.

Olsen, BE, Haugland, SA, Karlsen, E and Husöy, GJ (2005). Governance of complex procurements in the oil and gas industry. *Journal of Purchasing & Supply Management*, 11, 1–13.

Pisano, GP (1997). *The development factory: Unlocking the potential of process innovation*. Boston, MA: Harvard Business School.

Rainbird, M (2004). Demand and supply chains: the value catalyst. *International Journal of Physical Distribution & Logistics Management*, 34, 230–250.

Reichstein, T and Salter, A (2006). Investigating the sources of process innovation among UK manufacturing firms. *Industrial and Corporate Change*, 15, 653–682.

Reisman, A (1989). Technology transfer: a taxonomic view. *Technology Transfer*, Summer–Fall, 14, 31–36.

Reisman, A and Zhao, L (1991). A taxonomy of technology transfer transaction types. *Technology Transfer*, July, 38–42.

Rogers, EM (1983). *Diffusion of Innovations*. New York: The Free Press.

Roussel, PA, Saad, KN and Erickson, TJ (1991). *Third Generation R&D*. Boston, MA: Harvard Business School Press.

Rutherford, P and Persard, W (2003). Consider dynamic simulation tools when planning new plant startup. *Hydrocarbon Processing*, 82(10), 75–78.

Seal, RL and Turner, AL (1982). Design and installation of comminution circuits. In *Operator training and startup for semiautogenous grinding circuits*. Littleton: Society of Mining Engineers.

Seewald, N and Sim, PH (2003). R&D and customer collaboration key to boosting margins. *Chemical Week*, 165, 55.

Singh, K and Mitchell, W (1996). Precarious collaboration: business survival after partners shut down or form new partnerships. *Strategic Management Journal*, 17, 99–115.

Stewart, CT, Jr. (1987). Technology transfer vs. diffusion. *Journal of Technology Transfer*, 12, 71–78.

Stoneman, P and Kwon, M-J (1994). The diffusion of multiple process technologies. *The Economic Journal*, 104, 420–431.

Sveiby, KE (2001). A knowledge-based theory of the firm to guide strategy formulation. *Journal of Intellectual Capital*, 2, 344–358.

Thorp, B (1997). Mill/supplier relationships: the good, bad and ugly. *PIMA's Papermaker*, 79, 44–47.

Trott, P, Cordey-Hayes, M and Seaton, RAF (1995). Inward technology transfer as an interactive process. *Technovation*, 15, 25–43.

Utterback, JM (1974). Innovation and the diffusion of technology. *Science*, 183, 620–626.

Utterback, JM (1994). *Mastering the Dynamics of Innovation: How Companies Can Seize Opportunities in the Face of Technological Change*. Boston, MA: Harvard Business School Press.

Walters, D (2006). Demand chain effectiveness-supply chain efficiencies. *Journal of Enterprise Information Management*, 19, 246–261.

Verworn, B (2006). How German measurement and control firms integrate market and technological knowledge into the front end of new product development. *International Journal of Technology Management*, 34, 379–389.

Von Hippel, E (1986). Lead users: a source of novel product concepts. *Management Science*, 32, 791–805.

PART 5

PROCESS INNOVATION PERFORMANCE

In this final part of the book, the performance of process innovation will be discussed. Such a discussion must however be undertaken within the larger framework of total company innovation performance because of the intertwinement and complexity of innovation activities in the process industries. Because of that, each chapter starts with a presentation of the subject matter for innovation performance in general, followed by a discussion of process innovation performance in particular. It will be demonstrated that the key to fruitful development of success factors and performance measures is a breakdown of innovation into the previously presented different kinds of innovation activities as well as into hierarchic units. Finding success factors and performance measures for corporate innovation in general is thus a waste of energy and just as futile as trying to find a single measurable criterion of "beauty" (Eneroth, 2005).

Innovation-related activities for the company's internal customers such as "internal technical support" and "industrialization" will also be dealt with, since they are both often intimately integrated and carried out within the same part of an R&D organization. But not only that, the performance of application development will be touched on

267

briefly because it is in its inherent character similar to process innovation, but as such mainly process development carried out at the customer's premises. The sections referring to process innovation in this part of the book are mainly based on the author's original research and to some extent on information from other scientific publications. Potential success factors or performance measures of innovation-related activities and application development are however to a large extent based on the author's own industrial experience, because publications in those areas are non-existent.

Efficient implementation and use of success factors and innovation performance measurements rely on the development of a trustworthy and well communicated system for such measurements. Because of that, necessary time and company resources must be allocated to this development in order to be successful, and it is important to consider several aspects that have been well articulated by Packer (1983):

- The information must be understandable and easy to interpret.
- The information must be relevant.
- The information must be reliable and verifiable by others.
- The system must be accepted in the organization.
- The system must be cost-efficient.

The author's own practical experience of the development of success factors and performance measures for R&D has always proved to be a rather rewarding company exercise, and such an effort should therefore not be looked upon as a boring endeavour trying to satisfy external "share pushers", head-office "bean counters" and "innovation bureaucrats", but an interesting journey in innovation management, learning and change.

Chapter 13

Modelling R&D and Innovation Performance

"Company and R&D performance is likely to gradually decline if you do not carefully monitor and measure its performance. Once a slack management has allowed an R&D organisation to underperform, it is not easy to revitalise and bring back into good shape. As the prophetess says in Virgil's Aeneid (Virgil p. 134): The descent to hell is easy ... but to retrace your steps and escape to the upper air, that is the task, that is labour."

Thomas Lager

In times of ever-increasing demands on company performance, the R&D department has to address this issue as well. Shareholders' increasing focus on short-term profitability also puts pressure on company investment in R&D, which like all other departments must demonstrate improved development behaviour, showing that invested money is giving good payback to the company and is increasing shareholder value. As one executive phrased the question: "What bang do we get for the buck?" (Robb, 1991). So, as an allegory to the above quotation one should not let the company's R&D performance decline for lack of some kind of performance measuring system. It is however easy to ask for improved R&D performance, but it is not so simple to find the road to follow.

Innovation experts are often asked whether there is a positive correlation between business success and the amount of resources allocated

269

to innovation (R&D intensity). It is a well known fact that investments in fixed assets are important and necessary for long-term corporate survival, but there may be a negative correlation if there is overinvestment or investment in outdated products and production technology. The point is that the invested money must give a good return and thus be used in an effective and efficient manner. The same applies to innovation. There is not a positive correlation between R&D intensity and business success (Foster and Kaplan, 2001), because we have two hidden variables: R&D efficiency and R&D effectiveness. Pouring more resources into innovation without an effective strategy and an efficient R&D operation is a sure road to losing money, a fact that is also relevant to process innovation.

In Part 2, the development of an innovation strategy was considered to be the most important instrument for improving the effectiveness of company innovation. Part 3 then gave some tools for proper execution of such a strategy in the form of organizational structures and lean innovation work processes. Further on in Part 4, a collaborative open mode of development was recommended in order to improve the efficiency of R&D operations. Here in Part 5, Chapter 1 presents a simplified model of innovation followed by an introduction on how to develop success factors for innovation and process innovation in particular. The following chapter will then describe how the output from such successful R&D behavior can be measured, and the final chapter will show   how a measurement and monitoring system can be designed for innovation performance follow-up as well as for monitoring the implementation of strategic innovation intent.

13.1 A system perspective on innovation and innovation performance

In the first chapter of this book it was argued that there is a need for better theories of innovation as a guide to action for both academics and industry professionals. Figure 13.1 shows a simplified model of a firm's innovation which can be used as a "mental map" and serve as a general guide to understanding how improved innovation performance

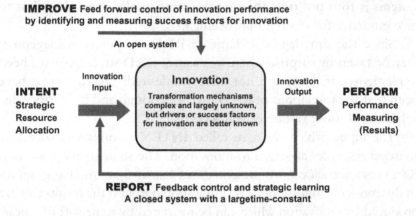

IMPROVE Feed forward control of innovation performance by identifying and measuring success factors for innovation

An open system

INTENT
Strategic
Resource
Allocation

Innovation Input

Innovation
Transformation mechanisms complex and largely unknown, but drivers or success factors for innovation are better known

Innovation Output

PERFORM
Performance
Measuring
(Results)

REPORT Feedback control and strategic learning
A closed system with a largetime-constant

Figure 13.1 A system view of innovation. The different parts of the system are named and illustrated in bold fonts, and in the last chapter they will also be used as names in the performance measuring system.

can be achieved. As such we will use Neelamkavil's definition of a system model (Neelamkavil, 1987):

> "A model is a simplified representation of a system (or process or theory) intended to enhance our ability to understand, predict and possibly control the behaviour of the system."

In Figure 13.1 innovation is simplified with an "input-output" model, in a reductionistic manner, when the innovation activities and transformation mechanisms for R&D are considered only as a "black box". In the perspective of such a theoretical model, the application and use of selected firm important success factors can then also be looked upon as "boundary conditions for good innovation behaviour", and as such necessary preconditions for the R&D model to be valid.

It is well-known that it is difficult to improve an industrial production process if the output is not measurable. Fortunately, the time constants in a production system in the process industries are often small, sometimes only in the range of days or even hours. Output measures are then very useful information, not only to control the process but also to improve it. One of the objectives of measurement systems is thus not only the measurement results themselves, but to give guidance for desired improvements.

Since the time-lag is so large in R&D, corporate management may be taken by surprise when, years after R&D strategies have been implemented the output is not what was desired! The cause may have been that the total input of R&D was not enough and the further distribution of those resources was incorrect.

The input, which we have called INTENT, consists of corporate strategic resources allocated to innovation. The strategic tools for such R&D resource allocations presented in Chapter 5 are the input figures to this model of innovation. The innovation output is the results that are produced by innovation which can be measured by some sort of "measuring instrument" or system which we have called PERFORM. There is however often a clear time-lag between input of resources and the realized output or outcome. In process engineering we talk about the time constant for a process, and if you want to control a complex industrial process with a large time constant you can often expect trouble. This is also the case when measuring the performance of a company's R&D.

In the old days when the voyages of ships could last several years, the proper reporting could bi- or even tri-annual. The annual company reports (not quarterly reports) are today, however, considered as a suitable span of time for measuring corporate performance, since business transactions can normally be completed over a year. Product and process innovation are, on the contrary, work processes that extend over a much longer span of time because of their large time constants and projects tend to span over several years, e.g. in the pharmaceutical industry, possibly up to ten years.

When ideas have been created and development projects formulated, it often takes time until projects are given company clearance to start up. Depending on the character of the development project, it may take one to five years before development is completed. In the case of product development, the products must then be launched on the market. For process development, and sometimes also product development, it may be necessary to invest in new plant equipment or possibly even whole new production plants. When those activities have been completed it takes some time to find out whether the development has been successful, and certainly an even longer span of time until the final economic outcome can be calculated. The problem with measuring product and process innovation can be formulated as trying to correlate today's innovation performance and output (the dependent variables) with yesterday's input of innovation strategy and innovation behaviour (the independent variables).

Because of the long time constant in the innovation system one cannot rely only on traditional feedback control using an analogy from process engineering control: a supplementary "feed-forward" approach is also needed to compensate for the time constant. We have to rely on knowledge of what we now generally call "best practice" or "success factors", which is called "IMPROVE" in Figure 13.1. The feedback from the measurements, which has been called strategic learning, is called "REPORT". This is important information, enabling corporate R&D management to understand whether selected strategic resource allocation has been successful or not, and will provide guidelines for strategy adjustments.

13.2 Financial or non-financial measures of innovation — the Balanced Scorecard revisited

The trivial fact that one can measure the economic performance of the past but not of the future is often overlooked by hard-core "you get what you measure" people. One definitely has a problem here, since companies' stakeholders certainly are (or should be) more interested in the future than in the past. For company employees, and certainly for the management, it should be more interesting to acquire some sort of understanding of how to behave successfully for the future, a kind of conduct which is sometimes referred to as "best practice", sometimes as "drivers" and sometimes as "success factors".

Three articles by Kaplan & Norton in the Harvard Business Review formed the backbone of the Balanced Scorecard, which is a song in praise of non-financial measures of company performance (Kaplan and Norton, 1992, 1993, 1996b). Nevertheless it must be observed that it is stated that the Balanced Scorecard complements financial measures of past performance with measures of drivers of future performance, thus emphasising in a natural way the starting point for the scorecard development and that measures to be developed are leading indicators. The strategy focus and the various measures to be developed are well summarized in Kaplan and Norton (1996a, p. 10):

> "The Balanced Scorecard should translate a business unit's mission and strategy into tangible objectives and measures. The measures represent a balance between external measures for shareholders and customers, and internal measures of critical business processes, innovation, learning and growth. The measures are balanced between outcome measures — the results from past efforts — and measures that drive future performance. And the scorecard is balanced between objectives that are easily quantified outcome measures and subjective, somewhat judgmental, performance drivers of the outcome measures."

The targets that are developed in a corporate scorecard are thus good directives for work process improvements or even re-engineering. After, and preferably even before implementation, the performance that relates to those success factors should be measured. Lacking a good system for performance measurement and feedback, any improvement system is likely to fail, as stated in the introductory citation. Thus, not only must the development of success factors be related to the firm context, but the measurements for R&D themselves can be developed and constructed along similar lines. The balanced scorecard is constructed from an overall company perspective (even if including innovation), but a rather similar perspective can also be applied to R&D.

There is often a consensus among R&D managers and company representatives that resources allocated to R&D can and should be regarded as investments. You allocate resources and expect something in return: an increased value of something such as a product or a company production technology (this could be denominated as "return on innovation" — ROIn). The paradox is however that this perspective often tends to be forgotten in the treatment of the annual company budget for R&D, when R&D expenses are normally treated as cost items.

The cause of this phenomenon is likely to be found in the often poor accounting for R&D performance, especially in the measuring of the return on R&D investments. Nowadays, progressive top managements often talk about R&D in terms of investments, although their follow-up of the intangible investments in innovation is unfortunately often poor or even non-existent.

There are however some fundamental problems associated with the treatment of R&D as an investment. The problems are mainly related to two issues: the ability to estimate the risk associated with innovation, and the problem of measuring the financial return from investments in R&D. It must also be observed that investments in R&D are also of two kinds; revenue-increasing and cost-reducing. Using the definitions of R&D given earlier, investments in product development usually focus on increasing revenues while investments in process development often have a focus on cost reduction. Looking at investments in R&D from a cash-flow perspective, it is also important to remember that successful R&D projects of either kind often call for further investment in production facilities and/or marketing and sales investments to reap the fruits of the R&D work, a fact well recognized and treated by Foster (1986).

13.3 Success factors or performance measures — where do we start?

The difference between "success factors" and "performance measures" is sometimes rather difficult to grasp. Before we go into the presentation of those concepts and their working definitions to be used in this book and the further introduction of success factors and performance measures of process innovation, you will be told a fable.

The fable of the overweight R&D manager

Once upon a time there was an R&D manager named Sebastian who, like many other people, was always struggling with his excess weight. He recalled a saying by one distinguished author that "you must not allow yourself to become a captive of your corpulence", and consequently he decided to do something about it. Sebastian first of all started to use his bathroom scale every morning to check his weight regularly. He thought that it must be important to have a good measuring instrument, and since he was an experienced R&D manager he said to himself that the weight he recorded every morning could be called a "performance measure" or a "key performance indicator". The bathroom scale was the measuring instrument. During his daily weighing exercises, however, he realized sadly that this did not help at all. He could only watch the unfortunate ongoing disaster on the bathroom scale.

Sebastian then began to contemplate the root cause of his overweight and soon came up with the conclusion that he was very likely eating too much and probably also the wrong kind of food. After reading a lot of books on how to eat in a different and healthier manner, he decided to drastically change his eating habits to find out if this was the solution to his overweight problems. From that day on he also started to keep a close watch on his daily calorie intake. After only a few weeks he found out that his weight started to go down steadily, and every time he stepped on the bathroom scale was a sheer joy.

As a trained scientist he immediately started to plot the results on a graph on which he had marked out a "target value" calculated by his body mass index. He started to call a close check on his calorie intake a "success factor" for losing weight.

During the early summer days he also noticed that his frequent golfing also seemed to affect his weight, and hence also recognized golfing as another "success factor" for reducing his weight. He even happily started to use his mathematical language and called the golfing and eating habits "independent variables" influencing the "dependent variable" weight, in a causal relationship.

By the end of the summer he had his weight fully under control, but unfortunately a small incident caused the bathroom scale to break down. To his astonishment he found out that he no longer needed the scale, but by measuring the coloured food points (he had now joined Weight Watchers) and keeping track of how many rounds of golf he played every week, he managed quite well to keep his weight under control. All of his neighbours noticed his successful slimming cure, and when his best friend asked him for advice he told him to join

Weight Watchers and take up golf again. He assured him that he did not even need a bathroom scale to check his weight daily, but measuring and setting up targets for those two previously mentioned success factors, which he also now called "leading indicators" was enough to assure his success. Sebastian explained that the weight recorded on the bathroom scale only gave evidence of what had already happened, and he therefore called it a "lagging indicator".

The moral of this story is that performance measures of R&D output are unfortunately measures of past performance, and furthermore do not give any guidance on how to improve R&D work and make it more efficient in the future. We must therefore scavenge for successful behaviour that we know from experience to be linked to better results. In other words we must look for "success factors" and good drivers for successful R&D. That is not to say that good "performance indicators" are unimportant; better to say that they are complementary measurables in the R&D performance improvement process.

13.4 Summing-up and some issues to reflect upon

Measuring R&D performance and productivity are important issues for management of R&D, and there have consequently been many publications in this area over the past two decades. Measuring R&D performance and productivity is, however, not very useful for company R&D management unless it gives an understanding of how well one is doing and how one could do better (Robb, 1991). This underlines the importance of relating measurement of R&D performance to R&D success factors. Some studies advocate a focus on R&D behaviour and success factors, while others stress the importance of output performance measures (Brown and Svenson, 1988). The answer is more likely that the two activities complement each other in an efficient process of total performance improvement for R&D.

In this chapter a system perspective on innovation and innovation performance has been introduced in order to explain the connection of strategic resource allocation to innovation with innovation performance and output. Lacking better models for the transformation mechanisms in the "innovation engine", success factors have therefore been introduced as a substitute and a condensed summary of good innovation behaviour for improved innovation efficiency (boundary conditions). The output from innovation is then given as feedback for not only an improved innovation strategy making but also for the improvement of success factors for innovation.

Further on, the similarity between the Balanced Scorecard approach at company level and the development of corresponding success factors and their measurables on an R&D level is recognized in order to demonstrate the need for the development and measuring of "leading indicators". Because of the difficulty of measuring some innovation activities in monetary terms, it is not easy to keep track of payback from innovation and especially from "innovation-related" activities. The balanced scorecard approach of measuring leading indicators is thus not only still relevant at company level, but even more so on a company R&D level, because of the long time constants in the innovation processes. This situation also makes the feedback control less applicable but makes "feed-forward" follow-up and control even more interesting for good innovation management.

The necessity for good "feed-forward" control and follow-up of company innovation stresses the importance of the development of the well-thought-out and solid innovation strategies discussed in Chapters 5 and 6. Such strategies must then be properly translated into distributed innovation intensities (and corresponding degrees of newness) for resource allocation at all levels in the company. This should be complemented by the development and measurement of relevant corporate-wide success factors for different kinds of innovation, an issue that will be further discussed in depth in Chapter 16.

- *Do you believe that measuring the performance of innovation and innovation-related activities is counterproductive from a perspective that "we should not disturb our creative people at R&D"?*
- *Referring to the model in Figure 13.1, how large do you estimate the average time constant is in your innovation system?*
- *Introduce this model and related control discussions to the process innovation staff and inquire if they find the analogy with process engineering relevant?*

- *Which of the two newly introduced indicators (leading indicator or lagging indicator) do you believe is most important for improving R&D performance?*
- *If innovation in your firm is unmeasured or measured by a few simplistic parameters like number of patents or research reports, what measurables would you suggest?*

Chapter 14

Success Factors for Process Innovation

"Fools learn from experience, wise men learn by other people's experience."

Otto von Bismarck

Measuring the performance of R&D has long been a priority for company executives and R&D managers (IRI, 2006). In the urgent need to develop and present performance measures for R&D, there is however a risk in picking some standard measures of output and performance indicators used by others without first reflecting on their use or whether the measures are actually relevant to their company's specific type of R&D. The R&D organization might reasonably take a sceptical view of such an introduction of ad-hoc measures of R&D performance.

A more sensible approach is to start with the question "How can the performance of our R&D be improved?" Measuring the performance of R&D is an important and worthwhile task, improving the performance of R&D is even more valuable. The R&D organization must first of all find out what is influencing its performance. Assuming that different kinds of behaviour lead to different degrees of success, it should be possible to find best practice or success factors that will improve development results, if well implemented.

The second word in the term "best practice" implies some sort of behaviour in the sense of "habitual or customary performance" (Webster, 1989). We then add the word "best" and imply that there is some sort of practice that is superior to other practices, evidently not only better, but the best. Using "science and best practice" is sometimes a recommendation for good professional behaviour in medicine, and best practice is then associated with behaviour that has proven successful over a long period of time (Lindahl and Lindwall, 1978). The term "best practice" is quite close to the concept of "success factors", which will be defined in the following section. The wording of both terms carries a risk, however, in that they can be interpreted to mean that a best practice for something is of an ever-lasting nature. History often tells us that a given kind of behaviour, and success related to that behaviour, is not likely to stay the same indefinitely: what is best practice today may be malpractice tomorrow.

In this chapter we will discuss the development of success factors. These can be looked upon as guidelines for good behaviour which hopefully, if implemented and followed, will improve the efficiency of an organization. Because of this they are also called "leading indicators". Good innovation work processes, a topic thoroughly treated previously in Chapters 8 and 9, have such already built-in relevant success factors in their design.

14.1   Success factors for process innovation — the drivers for the future

Good behaviour and best practice are supposed to be condensed into a number of well-defined success factors, but we may first of all ask the fundamental question whether it is possible to find "success factors" at all, and secondly whether the concept is useful for performance improvements of innovation in industry and process innovation in particular.

Good behaviour or best practice condensed into a number of success factors

The answer to the previous question must be of a more philosophical nature, related to the overall question of whether it is possible at all to find good practice in general and good management practice in particular. In the author's opinion it is similar to the question of whether it is possible to find good working practice for carpenters making furniture or for farmers in agricultural work, where the answer is "yes". It is however important to bear in mind the contextual and time dependence of best practice and success factors: making furniture for IKEA, for example, does not call for the same type of best practice as making 18th century cabinets.

If we then assume that it is possible to find best practice and success factors, do we need the concepts? In the author's opinion, the answer is "yes"! Before we start the discussion of indicators and measures of performance, we should focus on what we want to improve. Building a company system for performance measurement without a prior clarification of what one would like to improve is probably a fairly pointless exercise. The following definition is used here and was also presented to the respondents in the European research project, which we will return to in the following section (Lager, 2001):

> "In this context, we define 'success factors' in process development work as specific working methods and practices that lead to successful development projects (often called 'Best Practice' projects). The measure of success in development work and individual development projects may of course vary. In your assessment of the following potential success factors we would like you first to consider the importance of the selected success factor to future payback from the development, and then the importance of completing development on time at acceptable cost and meeting specified functional and quality criteria. Whether all these criteria for success need to be satisfied, or only some of them, naturally depends on the nature of the process development work."

In what follows here all factors that are or could be related to success are called "success factors". The degree of uncertainty, or how well they are proven, is expressed by the prefix "candidate" or "potential" for not proven, and "critical" for success factors that are well proven.

Success factors for what and for whom?

The problem of how to find and develop company success factors is related to the question of whether it is possible to learn from the behaviour of other companies and academic management research, or whether each company must develop its own specific success factors totally independent of external findings, in a never-ending iterative process of internal trial and error. If we assume that a company can learn from others, is there a risk in using success factors from, for example, other manufacturing industry in the process industries or in using the same success factors for process development as for the development of new products? The question is thus not so much whether it is possible to find success factors, but how contextually dependent those success factors are for different types of companies and for different kinds of R&D.

The complexity of company life, and especially the work in the R&D department, makes it important to consider the contextual dependence of success factors. Results from the Product Development Management Association (PDMA) success measurement project for product development indicated not only that product development success should be measured on two levels (project and company), but that the criteria selected should depend on what type of product development projects are carried out (Griffin and Page, 1991).

In a survey of publications on factors for success in R&D projects and new product innovation, the conclusion is that the literature shows many success factors for R&D, but which of these are relevant to different development contexts is unclear, and the notion of finding success factors of general applicability to R&D is considered naive (Balachandra and Friar, 1997). The need for a better classification system for success factors is noted in other studies. Figure 14.1 shows a classification system for the process industries and a tentative structure of success factors on different company hierarchic levels (Lager and Hörte, 2005).

The contextual dependence of success factors relates to the type of company activity to which the success factors apply. The activities of the R&D department have been tentatively classified into the areas previously defined in Chapter 3. Further categorization of those activities should depend on the type of company R&D, but it is suggested that the activities should be categorized at a lower level, for example differentiating between development that is more incremental and that which is of a more innovative nature (Lager and Hörte, 2005). The task specificity is shown on the left side of Figure 14.1. It is reasonable

Innovation & innovation related activities at the R&D function in the Process Industries	Success factors	Generic success factors	Sector-specific success factors	Company-specific success factors	Project-specific success factors
Internal customer	Development of captive raw material		▓		
	Development with raw material suppliers			▓	
	Process development	▓			
	Industrialization	▓			
	Internal technical services & support	▓			
External customer	Product development		▓		
	Application development			▓	
	External customer services and support			▓	
	Applied research		▓		
	Basic research	▓			

Figure 14.1   A classification and typology of success factors for innovation in the process industries. Success factors presented in the following section are shown in bold font. The shaded areas in the matrix represent areas where relevant success factors are likely to be found for different kinds of innovation. After Lager & Hörte (2005).

to assume that different success factors are more or less company-specific, so from a company perspective the question may be expressed as: "To what extent are certain success factors from other studies relevant to my company?" Figure 14.1 thus indicates where success factors can be located. Success factors for process innovation are likely to be of a more generic character, while those for application development probably are more company-specific.

Following this line of thought, there may be success factors that are generic, but not trivial, i.e. valid not only at company or sectoral level but even for all types of manufacturing industry. Success factors at industry level are those that are not company-specific but could be of use for many different types of companies in different sectors of industry. Company-specific success factors are no less important, but they must be developed within the organization. Finally, project-specific success factors must be considered in each individual project, since it is likely that some projects have certain specific circumstances that are unique. In Figure. 14.1 the organizational specificity is illustrated at the top, and success factors can be classified into those four categories but should not necessarily be structured in this dimension in a company's own system.

14.2   Critical success factors for process innovation— empirical evidence from the European research project

A study by Cooper and Kleinschmidt (1995) gives success factors for product development at company level (about 30 percent process industry in the sample), and the selected success factors in their study are more related to product strategy, the market and the product development process.

Out of the 72 candidate success factors presented by Balachandra and Friar (1997), those for R&D projects and new product development are strongly market-oriented. Consequently, success factors for process innovation are likely to be of a different character, which also will be shown in the following section and in the presentation of the research results. The framework presented by Balachandra & Friar, giving relative importance to different success factors as a consequence of the contextual environment for the R&D project, agrees with the approach in that study. It is evident that a listing of success factors is of little use by itself unless accompanied by a ranking or rating figure for each one, a fact that has been well recognized in this study.

The following results are from the previously presented European research project, which is further described in App. C. Six groups of industries were included in the study: mineral & mining, food & beverage, pulp & paper, chemical, basic metal industry and others. All success factors were generated by R&D managers from the process industries during face-to-face interviews. The success factors were categorized into five groups:

- Creating the development project.
- Working on the development project.
- Using the results.

- The internal environment.
- The external environment.

In the following text the importance rankings of the individual success factors are presented in parentheses (1 = highest ranking figure; 25 = lowest ranking figure; Increm. = ranking figure for incremental process innovation; Radical. = ranking figures for radical process innovation). Only some success factors have additional underlying explanatory success factors that were also generated during the interviews. The three highest-ranked of these explanatory success factors are presented for each success factor in ranking order.

Creating the development project

- The development organization is good at generating new ideas (Increm. = 15; Radical = 2).

  - Well formulated and clear technological visions that stimulate the imagination.
  - A clear and comprehensible problem description (preferably with the owner of the problem).
  - Development needs that are clearly expressed by the production organization.

The importance of a strong capability of generating new ideas and projects has been proven and discussed in several publications, for example von Hippel (1988). But how often are technological visions clear and well formulated in firms? Formulating a good technological vision is certainly the responsibility of the firm management in its technology planning, and of the R&D management in particular.

This is even more the case for process development, since "financing of long-term development work" was identified as an important success factor for process development. In a study of twenty projects in five companies, the most successful projects had a "project guiding vision" (Leonard-Barton, 1995). Formulating a problem well is often said to be the first step towards solving it (Hein, 1985); this is true not only in academic research, but evidently also in industry. This could be   something that each new project proposal or R&D project should put more focus on.

- Well formulated and measurable project objectives (Increm. = 1; Radical = 7).
- Clear definition of areas of technology in which process development work is to be performed (Increm. = 18; Radical = 15).
- Ability to identify and define "key surrounding issues" relevant to a project (for example price of energy, market conditions, etc.) and to relate them to project economy (Increm. = 12; Radical = 11).
- Ability to translate and to quantify an improved process economy into technical development targets (Increm. = 7; Radical = 13).
- Good risk analysis including a strategy in case of failure (Increm. = 25; Radical = 21).
- Well worked out preliminary studies with a clear interface to the following project phase (Increm. = 20; Radical = 14).
- A well structured project with clearly formulated and measurable "milestones" (Increm. = 8; Radical = 12).
- Frequent milestones that keep up the spirit and tempo in projects with long development cycles (Increm. = 17; Radical = 24).

Working on the development project

- There are good incentives and driving forces for process development (Increm. = 6; Radical = 5).
  - The project has the support of top management.
  - A strong project manager (champion) with the energy to pursue long-term projects.
  - The project has high priority among other development projects.

Many studies of industrial excellence have emphasized the importance of top management support. Other studies such as Roberts (1995) argue strongly for the importance of top management support for product development. The point is often stressed to such an extent that one sometimes wonders whether anything at all can be accomplished without direct support from top management.

Do we need a change of culture? We can now add process development to the long list of areas that need top management support. The second underlying success factor includes two important aspects: not only the need for a champion, but for one who can pursue long-term projects. Is the combination of top management support and a champion a winning formula for driving process development projects too? In smaller projects, a single factor could be enough to supply sufficient driving force. The results fit very well with the results from previous research.

- The total project group (or at least the core members) can be kept together during the total lifetime of the project (Increm. = 22; Radical = 18).

- The project group has a good and balanced composition (Increm. = 11; Radical = 10).

  - The project group has a good balance between theoretical and practical people.
  - Representatives from the production organization are included.
  - The project group has the ability to handle innovators and strong personalities.

The two highest-ranked underlying success factors state that the project group should include practical people (if we assume that production people are often gifted in practical matters). The highest-ranked of the three underlying success factors also stresses that the project group should include theoretical people. How often do project groups for process development include people with good theoretical knowledge? If the production organization is responsible for process development, it is unlikely that there will be so many people with a strong theoretical background in project groups. Conversely, if process development is in the province of R&D, there may not be so many practically oriented people. If we consider the case study of the Chaparral Steel mini-mill, one characteristic is that there is a natural mix of practical and theoretical people (Leonard-Barton, 1992). The need to have a mix of practical and theoretical people in the technology transfer process is also highlighted by Knight (1984).

- A well functioning and strong steering committee with the ability to pose difficult and important questions instead of just saying "yes" or "no" (a sound scepticism) (Increm. = 19; Radical = 9).
- Effective monitoring of the project and its results from start to finish (Increm. = 10; Radical = 17).

  - Sufficient time spent on consideration and analysis at the decision points for milestones.
  - Careful monitoring of "key surrounding issues" (including new ones) during the course of the project.
  - Detailed analysis of difficulties and failures during the project lifetime.

Project analysis at milestones is a main area of responsibility of the steering committee. How often do steering committee meetings allow enough time for in-depth discussions, and how often is the accumulated experience of the steering committee utilized in project work? There is a notion that steering committee meetings should be short and focused strictly on decisions. But ought this to be the case for complex R&D projects? One can also reflect that possibly too much has been written about the project group and too little about steering committees. The conclusion here is that careful analysis of project results at the milestones is a success factor, if sufficient time is allocated!

- Ability to judge the best environment (laboratory, pilot plant, demonstration, operating plant) for the different phases of experimental work (Increm. = 23; Radical = 22).
- The project is well communicated to the surrounding environment (Increm. = 16; Radical = 19).

  - The project is well communicated in the internal organization (understandable and comprehensible).
  - The project is well communicated externally (opening new doors, attracting development partners).

In the rich literature on project management the communication of the project is often discussed in terms of different directions of communication (Briner *et al.*, 1990). The ranking of the importance of internal compared to external communication makes sense, as it underlines that process development is more an internal affair. In product development there are many communication interfaces between developer and customer, developer and other external project allies, etc., whereas in process development we are not concerned with a product launch but with an efficient internal technology transfer to production.

Using the results

- Development results are measurable (significant, reproducible), trustworthy and distinguishable from other "process noise" (Increm. = 5; Radical = 23).
- The results are presented "packaged" in an understandable manner and efficiently "sold" to the customer (production, top management) (Increm. = 13; Radical = 16).
- Changes in the production process resulting from successful process development work are fully accepted by the production organization (Increm. = 2; Radical = 25).

The internal environment

- The company has a good and stimulating climate for process development work (Increm. = 9; Radical = 1).

  - The company also finances long-term development work.
  - Process developers get recognition.
  - The company dares to embark on projects involving high risk.

What do we mean by a good climate for development work? Three underlying success factors may well be too few to tap the dimension of a good climate for process development. One point to consider is that this question about a good climate may have been better answered by the R&D staff than by the R&D managers. The short-termism of company management is often discussed nowadays, but the influence it has on a company's working climate, especially on long-term activities like R&D, is seldom seriously addressed by top management. Since this success factor puts the finger on the importance of a long-term R&D strategy for successful process development, it is a matter to reflect upon seriously. How do the research staff interpret all signals regarding fast payback, and is more visionary long-term thinking in R&D departments stifled in its infancy because of this? One might have supposed that the two remaining underlying success factors — recognition of process developers and willingness to undertake risky projects — should have ranked higher, because risk analysis is always preached. On the other hand, the low ranking is congruent with the attitude at the Chaparral Steel mini-mill, where risk-less projects are avoided because they do not have the potential to outperform the competition (Leonard-Barton, 1992).

- The development organization includes individuals with suitable personal qualifications for process development work (Increm. = 4; Radical = 3)

  - Ability to distinguish what is important from what is not.
  - Strong belief in the project.
  - A wish to try and test new things.

What is a suitable personality for R&D and process development in particular? What criteria are used in the recruitment process for the R&D staff? The underlying success factors are probably too few to capture this complex success factor, but they could be a good starting point for further research. The individual's professional knowledge is not considered here, but is an area that should not be neglected in further development of company success factors. The top-ranked success factor — ability to identify what is important — certainly merits further thought, particularly about the true nature of that ability. Is this an important personal qualification for process development in particular, and what kind of ability are we talking about more specifically? A strong belief in the project is an interesting aspect to consider when setting up new project teams. Do the project members really have a strong belief in their project? This is an area that merits further investigation in the context of R&D in general.

• Strong mutual trust exists between the development organization and the production organization in question (Increm. = 3; Radical = 20).

The external environment

• The development organization is good at creating and engaging in development collaborations and alliances (Increm. = 24; Radical = 8).

   ▪ Attractiveness as a partner for collaboration in the selected project.
   ▪ Well-defined roles and division of responsibilities in the collaborative project.
   ▪ A well-balanced share of risks between individual collaborative partners.

The clear case of the top-ranked success factor — being an attractive partner — also has a bearing on the second success factor, because well-defined but different roles may be success factors in collaboration and alliances in general.

• Good and well-functioning networks are available for research and technical development (Increm. = 21; Radical = 4).

   ▪ Machine manufacturers and equipment suppliers.
   ▪ Universities and institutes of technology.
   ▪ External development companies and research institutes.

For companies in the process industries, the importance of collaborating with equipment manufacturers is well proven in other studies, e.g. Hutcheson *et al.* (1996). It can be noted that the importance of strong collaboration with equipment suppliers was recognized especially by the food and beverage, and pulp and paper industries. The importance of such collaborative development with equipment manufacturers is strongly emphasized and was consequently treated in depth in Chapter 10. The importance of collaborating with research institutes is also a fact well known e.g. to professionals in the basic metal industry.

- The development organization has good knowledge of conditions in the industry and its external business environment (Increm. = 14; Radical = 6).
  - Good knowledge of production structure and market conditions in the industry.
  - Good knowledge of the external infrastructure for the process in question (raw materials, energy etc.)
  - Knowledge of competitors' production technology and production environment.

This subject is often discussed in relation to company systems for information search and retrieval (business intelligence). The character of the success factor and the underlying success factors are of a kind that make them more important for radical process development than for incremental development (Lager and Hörte, 2002).

Discussions and conclusions

The most striking result to emerge from the ranking of success factors is the difference between the rankings for incremental process innovation and radical process innovation. Only four factors are ranked among the "top ten" for both of these types of process innovation, and then the rankings are very different for two of them. Only one of the "top five" success factors is the same for incremental process innovation and radical process innovation. Three of the top-ranked success factors for radical process innovation are in the "external environment" group, while three of the success factors for incremental process innovation are in the "internal environment" group.

The highest-ranked success factor for incremental process innovation is "well formulated and measurable project objectives", while the highest-ranked factor for radical process innovation is "the company has a stimulating climate for process development work". The second-highest-ranked success factor for incremental process innovation is "changes in the production process resulting from successful process development work are fully accepted by the production organization", while the second-highest-ranked success factor for radical process innovation is "the development organization is good at generating new ideas and formulating interesting new process development projects".

The results of the rankings for different types of process development clearly show that success factors for incremental process innovation and radical process innovation are different. Good project management is of great importance in incremental process innovation, which is often a plannable and optimization-oriented activity, and the ranking of success factors clearly reflects this. It is also reasonable to assume that well-functioning communications with the production organization are more important for incremental process innovation than for radical process innovation, where the working climate and strong potential to generate new ideas are likely to be of greater importance.

The difference in ranking between incremental process improvement and radical process innovation seems to be strongly related to the difference in character between the different types of process development work. The difference between success factors for incremental and radical process innovation proves that there is a need to better distinguish between process development work of different nature and content, and a need for a better classification of process development projects, a fact that has been discussed in depth in Chapter 4. It is evident from the results that simply listing success factors is of little use unless a ranking or rating figure is assigned to each individual factor. The findings in this study agree well with the framework presented by Balachandra and Friar (1997), giving relative importance to different success factors as a consequence of the contextual environment for R&D projects.

There was generally a very good agreement between the six categories of industries (Lager and Hörte, 2005). The three least important success factors were rated among the least five important in all six categories. Similarly, the two most important success factors were ranked among the top five by all six categories.

One can say that the overall impression of the results is that there is good agreement between different categories of process industry. This supports the general idea of this book as treating the process industries as a cluster with many family resemblances. It must also be observed that a high rating of a success factor in an area where the company is currently performing well does not necessarily imply that the company needs to do even better in that area in the future and over-perform — an issue that has been pointed out by Leonard-Barton (1995).

Success factors and their individual rankings are of interest in themselves, but the pattern that emerges from the average ranking points for the different groups of success factors is even more interesting. "Using results from process development (technology transfer)" received by far the highest ranking points. Sometimes industrial development is likened to a funnel, and it is often argued that there is a need to pour many new ideas into the funnel of the development process to get a few good usable products or processes out at the bottom. This may give the impression that this phase of the development cycle is the most important. It is interesting to note that the results of the present study suggest that the technology transfer phase is very important in the process development process, a fact that was well recognized in the development of the process innovation work process presented in Chapter 9. The conclusion is not of course that there is no need for many good ideas in the beginning, but that one also needs to understand how to make the technology transfer more reliable in process development, a fact that was highlighted in the treatment of technical transfer in Chapter 12. It must also be observed that technology transfer is not the sole responsibility of the R&D department; the production department also needs to develop a good receiving capacity for new technology.

In three of the ten top-ranked success factors the theme is trust; the results are "fully accepted", "trustworthy results", strong "mutual trust". How do we create a climate of trust for the use of the results? This brings us close to the question of user involvement in process development, a question that was not identified explicitly during the interviews but is strongly related to how process innovation is organized, (see Chapter 7). User involvement is an interesting question that so far has not been given a clear answer, but simplistic treatment of the concept may lead to unclear results that miss the points of user selection, timing of user involvement, the user's willingness to be involved, etc. (Leonard-Barton and Sinha, 1993).

14.3 Potential success factors for industrialization, internal technical support and application development

The previous presentation of success factors for process innovation was based on solid empirical evidence from the European research project. Lacking similar evidence for industrialization, internal technical support and application development from academic publications, a number of potential success factors for those areas have been tentatively compiled from industrial contacts and seminars with industry professionals. Some new potential success factors for process innovation collected after the European research project using another structural grouping have also been added. Empirical evaluation and verification of the following potential success factors will be welcome.

Supplementary potential success factors for process innovation (not in ranking order)

In the previous section, reported success factors for process innovation from the European research project were structured, with the assistance of industry representatives, along different phases of a project's life cycle.

Those results are presumably still valid, but later on additional information of a more ad hoc character has made it of interest to add a few more potential success factors as well as presenting them in a slightly different structure. Neither of the two lists is superior and it is possible for firms to choose among both "menus" in order to seek inspiration and to select success factors more in an "à la carte" manner. It is also desirable for individual firms to add and develop more company-contextual success factors for process innovation.

A deep knowledge of the company's own and competitor's
process technology

- Visits to competitors' production facilities (if possible).
- Good knowledge of process innovations from attending conferences.
- Collaboration with non-competing firms in both own and other industrial sectors (strategic partners for collaboration).
- Solid knowledge of the company's production process capabilities.
- Focusing on core process technology development.
- Good knowledge of factors that tend to disturb the company's production processes.
- A holistic view of the company's total production chain.
- Good understanding of the relationship between product specifications and production process parameters (an ability to predict product specifications from process parameters).
- A thorough understanding of the relationship between product specifications and process control.
- Ability to find production bottlenecks and remedy them.

A well functioning process innovation work process

- A well-functioning and well-structured process innovation work process.
- Good integration between product innovation and process innovation teams in product innovation projects.
- Good knowledge of scale-up.
- Good collaboration with the internal customer.
- Easily accessible process developers.
- Regular meetings with plant managers about long-term process innovation needs.
- Ability to reformulate a process problem into a development project.
- Good implementation of new process technology together with production (good technology transfer).

Potential success factors for industrialization (*in no ranking order*)

- People with knowledge of industrialization should be on the steering committee.
- People with knowledge of industrialization should be in the project group.
- Clear and useful risk analysis for new investments.
- Solid scale-up calculations for new process technology.
- Representatives from both product and process development should be in the project group (securing technology transfer from R&D).

Potential success factors for internal technical support (*not in ranking order*)

In this book we distinguish between process innovation and internal technical support in the process industries. Internal technical support is not considered as an innovation activity but an "innovation-related activity". As such, the knowledge needed to solve a technical problem or to improve a process through technical support must then already be available at the R&D department "in-house". If solving an assignment calls for development of new knowledge, the assignment should be reclassified as a process innovation activity. This is of course only "strictly speaking", and neither party must behave in a bureaucratic manner. It is however felt that it is important to discuss and distinguish between both kinds of activities, since they are often carried out by the same people at the same R&D department, but with clearly different strategic objectives.

The following list of potential success factors for internal technical support to production is recommended for use only as a tentative starting point for a consecutive development of a company's list of its own, more contextually relevant success factors. Each success factor could then be benchmarked when the individual factors are ranked or rated on a scale, and the company's performance in each area evaluated.

Good knowledge of the company's own production

- Good knowledge of the company's own process technology.
- Good knowledge of the total cost structure of the company's process technology and especially of unit process operating costs.
- Expert process teams for problem-solving and trouble-shooting.
- Clear understanding by production of whom to contact for different kinds of assignments.

A good work process

- A named client for each assignment.
- Technical support personnel easily accessible to clients.
- Clients should be respectfully treated during the assignment.
- A clear and well formulated objective for each assignment (preferably jointly worked out together by both parties).
- Good early estimates of whether a desired time schedule is achievable (good knowledge of internal priorities and workloads).
- Professional and methodological problem-solving (alternative solutions).
- Close and frequent contact with the client during the assignment.

Trustful and well-functioning collaboration

- Informal day-to-day collaboration and communication.
- Technical support staff should be visible at production sites in between assignments.
- The potential client should never be let down by technical support if he/she is in urgent need of help.
- Straight talking and no "bullshit".
- Regular meetings when new process technology is introduced and discussed.

Good implementation of results

- The reward for technical support is the client's success, not your own!

- Implementation of the results with the client, hopefully with no process disturbance.
- Technology transfer and education for production personnel (striving to make technical support obsolete).

Potential success factors for application development (*not in ranking order*)

In this book, we make a distinction between product development and application development in the process industries (see Chapter 3, Figure 3.3). Application development is definitely considered as an innovation activity but the knowledge required to solve a problem or to improve a process at the customer's premises by application development will not need significant company own product development. Skill in application development is however strongly linked with a good knowledge of the company's own products in combination with an intimate understanding first of all of the customer's production processes.

The potential success factors for application development listed below should only be used as a tentative starting point for a company's development of a list of its own, more contextually relevant success factors. Figure 14.1 shows that success factors for application development are likely to be more company-specific than success factors for process innovation. Each success factor could then be benchmarked when the individual factors are ranked or rated on a scale and the company's performance in each area is evaluated.

Intimate knowledge of the customer's process and product technology

- Ability to select the proper form for contacts; a good speaking partner (meetings, seminars, etc.)
- Good knowledge of the company's own products and their applications (an intimate understanding of when the company's own product is functioning well or not so well for the customer).

An efficient "application development work process"

- Easy for customers to contact company representatives for application development.

- A transparent work process that has been communicated to the customer.
- Ability to select the proper form for development (more or less formal projects, trouble-shooting, etc.).
- Ability to reach a mutual agreement on where to conduct the development work (at own company's test facilities, customer's test facilities or other external test centres).
- Ability to select the cheapest and/or best test environment for the application development project (simulation, laboratory test work, pilot plant test work or tests in the customer's process plant).

Trustful well functioning collaboration

- No problem for personnel to visit the customer's production site (willingness to travel).
- Respect for the customer and the customer's arguments and point of view.
- Good command of the customer's language or preferred second language.
- Not afraid to make suggestions for solutions to problems to the customer (self assured).
- Regular application development meetings or seminars with the customer.

Good implementation of results

- Flexibility in working hours with the customer.
- Ability to estimate the process benefits for the customer; "value in use".
- Understanding how the company's own product will function together with the customer's other raw materials or commodities.

14.4 Summing-up and some issues to reflect upon

In this chapter the concept "success factor" is introduced and it is first of all discussed how firm-specific such success factors are. Further on success factors for different kinds of innovation activities are presented. Success factors for process innovation are introduced and it is shown that such success factors are different for incremental and radical process innovation. Lacking empirical evidence, potential success factors are presented for industrialization, internal technical support and application development.

The strength of nurturing core capabilities may have the opposite effect and turn them into core rigidities. Core rigidities are thus "good old core capabilities" that are no longer important and may even be harmful to the company in the future. There are several examples of such cases; the subject has been thoroughly treated by Leonard-Barton (1995).

Updating the importance rating of success factors is one way to prevent core capabilities from becoming core rigidities. Not only should the importance ratings of already selected success factors and the targets be adjusted, but obsolete success factors should be removed and new ones added.

Making this revision on a regular basis will make the success factors dynamic and help to achieve the gradual evolutionary adjustment that is recommended as preferable to revolutionary change (Leonard-Barton, 1995). Success factors should be fed back to the work processes, and such well-functioning work processes are then those where success factors are well "embedded" in the work process structure.

- *Do you believe that talent in management of innovation and technology is something one is born with, or must it be acquired in a process of learning by doing, and thus is something that cannot be learned from others?*
- *Is your R&D organization trying to capture good innovation management behaviour and skills and are you trying to accumulate this information as some sort of internal best practice (success factors) for innovation in an organizational learning ambition?*
- *Review the success factors presented for process innovation and sort them in ranking order of importance for your firm. Benchmark your firm's abilities in these areas.*

- *Review the potential success factors for industrialization and sort them in ranking order of importance for your firm. Benchmark your firm's abilities in these areas.*
- *Review the potential success factors for internal technical support and sort them in ranking order of importance for your firm. Benchmark your firm's abilities in these areas.*
- *Review the potential success factors for application development and sort them in ranking order of importance for your firm. Benchmark your firm's abilities in these areas.*

Chapter 15

Key Performance Indicators for Process Innovation

"Not everything that counts can be counted; not everything that can be counted counts."

Albert Einstein

One could initially question whether there is any need at all to measure the output from R&D in the company. Opinions on the importance of measuring R&D may differ within the company, and certainly also within the R&D organization. This is an issue which was put rather well by a former R&D director of ICI America (Tipping, 1993):

"With R&D expenditures around total company profit, it is a significant temptation in bad times (The Board of Directors) — To those who believe, no explanation is necessary; to those who do not, no explanation is possible (The research community)."

The need for a more transparent estimate of how important R&D is to a company and how well it is performed is strongly advocated by Robb (1991), and measuring R&D is also recommended to be put high on the agenda for company R&D management. But, referring to the quotation above, is it possible to measure complex matters like company R&D, and if so do we have a tendency to measure what can be measured instead of measuring the important things? One is reminded of the old story about the man who lost his keys in the evening on the street and searched only under lamp-posts because it was easier to see there. The answer to the first question is yes, but the more complex and multifaceted an undertaking or phenomenon to be measured is, the more complex measuring instruments and number of measurables are often needed. This is however something that the R&D staff is well aware of, and since industrial R&D is usually a rather complex undertaking, this also applies to the measurement of that activity.

From a historic management perspective, the lack of relevant measuring and control systems for innovation is probably related to a rather old-fashioned view that such an activity is too complex to be measured. But the knowledge base in this area has greatly improved over the past few decades, and we now know vastly more about how to measure the performance of R&D (Ellis, 1997; IRI, 2006; Kerssens-van Drongelen, 1999). There is today a spectrum of alternative financial and non-financial performance measures that can be selected depending on the nature of company R&D and on which areas it is desired to improve. After success factors for innovation have been developed, it is the proper time to develop output measures for company innovation. The participation of company R&D staff in this process is essential. People in the process industries, often being experts on other kinds of measurements, are not likely to accept measures that are not understandable, relevant and reliable.

15.1 From a corporate black box to innovation transparency

Although all internal functional units are cost centres and not profit centres, there is one big difference between R&D and other functional units. If marketing and sales shuts down, the response and consequences will be immediate, as the firm will then cease to sell any products and will not earn any revenues. The outcome of marketing and sales can easily be measured, and the results are measured not only in numbers of products sold but can be directly aggregated in financial terms too. So be sure that marketing and sales results will be closely followed for better or worse.

The company's production department is always measured in detail in all dimensions, and since both input and output are easily measured, if not always in financial terms, productivity can also be calculated. Improving productivity in production has always been an issue of the greatest importance to any company that wants to stay cost-competitive in today's global markets. The need for good productivity measurements at all levels in the company is well put by Brinkerhoff and Dressler (1990 p. 9):

> "Common among almost all of the productivity improvement strategies in use is measurement. Measurement is used to indicate whether there is a need for any improvements in the first place, is often a part of the improvement process itself, and is used to gauge whether improvement efforts are making any progress. Often, in fact, measurement alone has a dramatic impact on productivity since the effects of feedback are so powerful."

Productivity in production can thus easily be calculated and often compared to similar firms and even competitors. If production plants are shut down, be sure that this will not pass unnoticed. The same goes for logistics and some other functional units. But what about R&D?

Measuring outcomes and output from R&D — evidence from the past

The R&D department delivers its output into the company's internal operating environment. But however efficient and effective product and process development work in the R&D department may be, if this output is not transformed into successful products on the market or improved production technologies, the final outcome of R&D is nil. This is further illustrated in Figure 15.1, where three different levels of analyses are presented for R&D; success factors and performance indicators can thus be found on all of those levels. The company, the R&D department and projects can be looked upon as transformation processes receiving input of various resources and delivering output in various

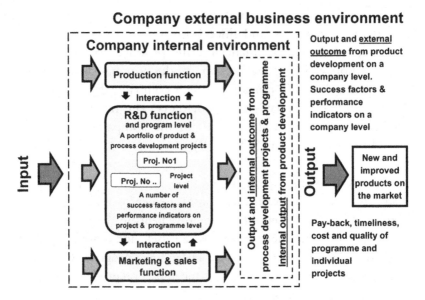

Figure 15.1 A conceptual model of R&D at different levels of analysis; project level, R&D or programme level, and company level (Lager and Hörte, 2002). The output from process innovation is internal, while the output from product innovation is external. The share in successful new products on the market accounted for by R&D is not easy to distinguish.

forms. In Figure 15.1, a distinction has been made between "output" as a deliverable in a subsequent chain of outputs, and "outcome" as a final deliverable or result that is measurable in an economic dimension. Process development is easier to measure in this respect than product development because it delivers its outcome within the company operational environment. Not only the importance of understanding successful R&D behaviour, but also the difficulty of measuring the share of R&D of the outcome from development work and the time lag between the start of R&D activities and future economic payback, make it interesting to try to find relevant success factors.

There is a time-lag in R&D that is sometimes forgotten, not only by the outside world but to some extent sometimes also by company management. But if one is aware of this time-lag, it could be taken into account to some extent in the calculations. In some companies it may even be possible to shut down R&D for a while without anybody

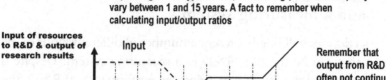

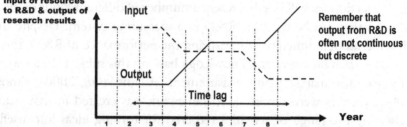

Figure 15.2 Input of company resources over time to R&D (dashed line) and R&D output (solid line). As a result of a time-lag of several years, the results from R&D are strongly lagging. As a consequence, at the end of Year 6, it looks as if a successive reduction of R&D expenditures has resulted in increased R&D output and high productivity.

noticing. Depending on the nature of the company's R&D activities, some areas of R&D will not Influence company results for many years. If a large pharmaceutical company shuts down its research departments, for instance, it might even take three to five years until the results show up in company revenues (see for example Figure 15.2). On the contrary, since R&D often consumes 10–20 percent of a pharmaceutical house's turnover, profits will rise. On a short-term basis, shutting down R&D would thus give a substantial increase in profits over several years.

To cut a long story short, the time constant for R&D, together with the usual lack of output measurables of financial return, makes R&D extremely vulnerable in times of low company profitability. R&D must prove its case better, and if it does not take up this challenge, there is a great risk that the merits of company R&D will not be recognized and gradual cuts in R&D budgets will be very difficult to spot in a short-term output perspective. It is however not company top management that should solve this problem. It is solely the responsibility of R&D management to constantly seek and develop good performance measures in order to demonstrate its value to the firm.

15.2   Exploring innovation performance measurables and a measuring system

In this section we will briefly present a number of different kinds of measurables that can be used for R&D in a measuring system. Despite the aforementioned difficulty of measuring the performance of R&D, there is today a rich literature and knowledge base on this subject. In a review of innovation management measurement (Adams *et al.*, 2006), almost 200 references were reported. A framework was created in that study "covering the range of activities required for turning ideas into useful and marketable products". A classification of the measurables was made into the categories of inputs management, knowledge management, innovation strategy, organizational culture and structure, portfolio management, project management and commercialization.

The idea of measuring different aspects of innovation and the innovation process is important, but the focus in that study is only on measuring product innovation and not process innovation. The importance of designing specific measurables for different kinds of R&D is also highlighted in a recent study of key performance indicators for research (Samsonova *et al.*, 2009). Measurables for process innovation and process-innovation-related activities are, however, scarce. This section will start with a brief presentation of different kinds of measurables, categorized according to a different dimension.

Qualitative measures

These measurables are often also called subjective or intuitive measures and their values are not expressed in figures. This kind of measurable is often used when a quantitative measurable is difficult to find, and as such often in more research-oriented R&D (Werner and Souder, 1997). The "peer review" of scientific activities is a good example of qualitative measures that should also be considered as a useful instrument for applied research in industry.

The "Jimmy Stuart test" is another example of a qualitative measurable that can be used where good alternative quantitative estimates are lacking.

"Jimmy Stuart tests" — A historic review of breakthrough innovations of large strategic value influencing the corporate body

Walter Robb (1991) has introduced this rather interesting qualitative measure which he calls the "Jimmy Stewart test". The story goes as follows:

> "The first is what I call the 'Jimmy Stewart test', after the actor. One of his most famous roles was in a movie called 'It's a Wonderful Life.' In it he plays a man who runs a small-town savings and loan. A sudden setback has left him discouraged and he is considering a suicide. At that point, his guardian angel comes down from heaven and figures out a way to cheer him up and save his life. The guardian angel shows Jimmy Stewart how much worse the world would have been if he hadn't lived. In much the same way, we can ask ourselves a similar question. What would our companies look like today if a central research laboratory had not existed? What business would the company not now be in? What profits would be forgone?"

This kind of qualitative measure of R&D output is to be regarded as supplementary, and the author's personal experience of practising this is very positive. In a session the R&D representatives can then be asked to pick the three most important R&D achievements during the past two decades and to present them as "mini-cases", including a rough estimate of their overall importance to the firm. Many process innovations often tend to appear in such a context.

Semi-qualitative measures

Pure qualitative measures are usually converted into what is called a "semi-qualitative" measure when the qualitative answer is translated into a number on a selected ordinal scale (Galtung, 1967). This group of measurables includes classical benchmarking activities, and their reliability depends on whether it is a self-assessment or if someone else in the organization or an outsider is doing the assessment. Selecting a proper ordinal scale for such measurements is the first step. A five-point scale (1–5) is often selected but if better resolution is desired a nine-point scale is often preferable (1–9). There are different pros and cons to giving a verbal statement of the meaning of the individual numbers on the scale.

Quantitative measures

The simplest kind of quantitative measures are often called "biblio-metric methods" (bean-counting). An example of such a measurable is counting "number of patents" or "number of scientific publications in refereed journals". Bean-counting can also be combined with other quantitative measures. Patent counting alone is not a very good measure, but if it is combined with the quantitative measurable of the cost of producing the patent, it gets better. Number of patents per megabuck (one million dollars) is one measurable that has been supported by GE Schenectady Laboratory (Robb, 1991). In addition, supplementary information about those patents like number of fields, foreign filings, licensees and outcomes makes the measuring of patents more relevant as a measurable.

Economic measurables

Assuming that both input and output can be translated into costs and profits, this is naturally a most desirable measurable, since it is possible to aggregate at different levels in the company. One problem with ratios of output to input measures, however, is that a high score can reflect a decrease in input rather than an increase in output.

Combinations of measurables into a measuring system

It is of course desirable to measure R&D performance like all other activities in terms of standard business financial measures such as profit, return on investment, etc., but Kerssens-van Drongelen and colleagues (2000) suggest:

- It is difficult if not impossible to isolate R&D's contribution to company performance from other business activities.

- It is problematic with today's accounting methods to identify R&D's contribution to the profits resulting from individual products.
- The time-lag between R&D and the potential financial rewards makes it difficult to use this information.

The solution to this problem is not to use a single measurable but a spectrum of measurables of different kinds. It is a well-known fact not only from economics but from natural science that the more complex the phenomenon to be measured, the more complex the measuring instrument or system needed. The results of a recent survey of R&D-intensive Italian companies not only show that there is a strong interest in employing such measuring systems for R&D, but that many companies have already been using them for a long time (Chiesa *et al.*, 2009a, 2009b).

In an early review of measuring R&D performance (Werner and Souder, 1997), over 90 articles were found describing various techniques. Integrated methods of R&D performance evaluation were recommended because they combine qualitative and quantitative measures, thereby enhancing the advantage of both types of measurements.

A combination of different measurables is thus an interesting solution, and Schumann *et al.* (1995) give a classification of different measurables, using the matrix presented in Figure 15.3. The four quadrants represent:

- Internal end-of-process measurements: those discussed in this chapter that can be used to keep track of output over time.
- Internal in-process measurements: more efficiency-oriented, but can also include measurables of success factors presented in the previous chapter.
- External in-process measurables: these include results from functional benchmarking when the competitive aspect is small.
- External end-of-process measurements: results from competitive benchmarking achieved in joint benchmarking exercises or from internal business intelligence.

Measurements in all four quadrants are important, but for different reasons.

	End of process measurements	In process measurements
Internal measurements	**Performance tracking**	**Technical productivity improvements**
External measurements	**Competitor assessments**	**Benchmarking of best practice**

Figure 15.3 Different types of R&D measurables. After Schumann *et al.* (1995). A typology has been developed using the two dichotomies for where the performance is measured. The recommended use of such measurables is presented in the matrix.

15.3 Potential key performance indicators for process innovation and innovation-related activities

Measurables for process innovation and innovation-related activities are scarce. In order to present some suggestions for in-house development of such measurables, a number of "potential measurables" are presented in this section that have not been verified in any scientific empirical study. They must therefore be regarded as propositions that need to be empirically tested. Output measures are shown in bold italic fonts, while measurables of a more success-factor-oriented character are shown only in italics.

A company trying to develop performance measures can use these lists of potential measurables in internal exercises both as success factors and as "triggers" for in-house development of more company-specific measurables in its systems for performance measurement.

Potential key performance indicators for process innovation

As stated in the caption to Figure 15.1, the advantage of measuring process innovation compared to product innovation is that process innovation delivers its output into the company's own internal environment. Process innovation measurables are therefore more easily related to quantitative or monetary measures.

Table 15.1 Potential Key Performance Indicators for Process Innovation.

To be measured	Unit of measurement	Direction for improvement
Aggregated yearly benefit for production* from process innovation (profitability of investing in process innovation)	Total € per year	The more the better
Aggregated process innovation project proposal quality (reliability of investment estimates)	percent of estimated potential	The more the better
Aggregated process innovation project cost (reliability of estimated investment cost)	percent of estimated budget	The lower the better
Aggregated process innovation project timeliness (reliability of time estimates)	percent of estimated schedule	The lower the better
Number of process patents per year (this depends on corporate patent strategy)	Number	Target value (depending on patent strategy)
Number of cross-functional collaboration project groups (ability to set up the optimal process development teams)	*Started during the year*	Target value
Number of people with production or operating experience in the process development organization	*Number*	Target value

* The aggregated yearly benefit to production can be estimated in the following manner: each individual development project should, like all innovation projects, have an estimated potential if it succeeds. When a project is finished and reported, the report should have a standardized "last page" where the client should give some feedback on the outcome of the project. This feedback should include an estimate of how much of the project potential has been reached. Such an evaluation should preferably be carried out jointly with the client and the project manager. All project results are then accumulated on a yearly basis. If the potential is not expressed in monetary units (e.g. increased throughput), those figures are to be recalculated in estimated monetary figures.

Table 15.2   Potential Key Performance Indicators for Technical Support.

To be measured	Unit of measurement	Direction for improvement
A yearly customer survey to all internal clients for customer support (mainly aimed at the production department but could also be relevant to others, and is preferably on-line)	The results	As good as possible
Number of started assignments for production per year (measuring some sort of attractiveness of this function; also whether there are too many assignments)	Total number	Target value (not too many if time is to be available for process development)
Number of finished assignments to production per year (observe a time delay)		As high as possible
Startup time after first contact with production (responsiveness to client's demands)	Average number of weeks	As low as possible
Number of people with operating experience of company processes (ability to communicate with production people and practical process knowledge)	Number	Target value (a proper mix of practical and theoretical people desired)
All assignments are individually evaluated by each customer	Average score	As high as possible

Potential key performance indicators for internal technical support

Measurables for internal technical support are also easier to construct because the services are delivered within the company. Appropriate measurables will have a more qualitative dimension which does not make them less valuable.

Table 15.3   Potential Key Performance Indicators for Application Development.

To be measured	Unit of measurement	Direction for improvement
Number of people with customer process operating experience or development experience with a customer (a measure of how many people in the company working on application development possess this knowledge)	Number	Target value
Number of application development projects initiated this year	Number of new projects	Target value
Number of application development projects with customers (attractiveness of the company for joint development)	Number of on-going projects	Target value
Number of application development projects finished this year with a "hit" (improved sales)	Number of successes	As high as possible
Number of people with end-user process operating experience or development experience (a measure of how many people in the company working on application development possess this knowledge)	Number	Target value
Number of collaboration projects including both customer and end-user (ability to create this kind of "super project")	Number	Target value

(*Continued*)

Table 15.3   (*Continued*)

To be measured	Unit of measurement	Direction for improvement
Improvement of "value in use" for the customer's product or process (This is the ultimate thing to measure for application development, but naturally not so easy to obtain from the customer)	€ estimated by the customer	As high as possible
An application part of the customer yearly survey	Average score	As good as possible

Potential key performance indicators for application development

Measurables for application development are more in the nature of measuring efficiency-oriented behaviour — something like to measuring success factors. Getting the customers to assign a value to their improvements may often be difficult, and the outcome of application development projects may thus be difficult to measure in economic terms.

Potential key performance indicators for industrialization

Industrialization is delivered to the company's own organization, which makes it easier to find suitable measurables.

15.4 Summing-up and some issues to reflect upon

This chapter discusses key performance indicators for innovation in the process industries, explaining why corporate innovation activities must be measured at lower hierarchical levels to achieve relevant measurables. This emphasizes the importance of good definitions of innovation activities. The more complex the undertaking to be

Table 15.4 Potential Key Performance Indicators for Industrialization.

To be measured	Unit of measurement	Direction for improvement
A yearly customer survey to all internal clients (general issues) (mainly aimed at the production department but could also be relevant to others; preferably held on-line)	Results	As good as possible
Average time for pre-studies of industrialization projects (Measuring the company's capacity to plan projects in advance; securing thoroughly prepared investments)	Months	Target value (six months might be proper for a good pre-study)
Benchmarking and evaluation of all industrialization projects (by steering committee, project group and the internal customer)	Results	As good as possible
Evaluation of plant (process technology) operating performance after one year (comparison to total project objectives)	Comparison to objectives	As high as possible (percent of objectives)
Reassessment of plant (process technology) operating performance after four years	Comparison to objectives	As high as possible (percent of objectives)
Start up efficiency (lead time from startup until project objectives are achieved)	Average months for all projects	As low as possible (comparison to project plans, must depend on the complexity of the technology)
Number of people with startup experience in the project group	Number	Target value

measured, the more complex the measuring instruments needed. The complexity of innovation in the process industries must therefore be considered when a suitable measuring system is developed. The time constant for all innovation activities is discussed, and a number of alternative measurables are suggested. It is concluded that using well-developed key performance indicators for innovation also gives opportunities for both internal and external benchmarking. Feedback from measuring innovation performance is important information for the innovation strategy-making process in order to assess the outcome of such strategies.

- *One can assume that the internal R&D controller keeps good track of R&D expenditures. On the other hand, how well is the profit side of R&D accounted for in the corporation?*
- *Has it ever been discussed in the R&D organization how output from R&D could be measured and how such figures ought to be communicated to the internal corporate stakeholders who allocate resources to R&D?*
- *Pick the three most important R&D achievements in your firm during the past two decades and present them as "mini-cases", including a rough estimate of their overall importance to the firm.*
- *Review the potential key performance indicators for process innovation and sort them in ranking order of relevance to your firm.*
- *Review the potential key performance indicators for internal technical support and sort them in ranking order of relevance to your firm.*
- *Review the potential key performance indicators for application development and sort them in ranking order of relevance to your firm.*
- *Review the potential key performance indicators for industrialization and sort them in ranking order of relevance to your firm.*

Chapter 16

A Corporate Follow-up System for Strategic Intent and Innovation Performance

"When you have assembled what you call your facts in logical order, it is like an oil-lamp you have fashioned, filled and trimmed; but which will shed no illumination unless first you light it."

Antoine De Saint-Exupéry, 1984

In large corporations in the process industries, resource allocation to R&D is nowadays often decentralized and investment in R&D, depending on the nature and scope of innovation activities, is often left to the discretion of business units, divisions, subsidiaries or even smaller production units. The advantage of this is that it ensures that innovation activities are in line with individual business strategies, that product development strategy is close to the external customers, and that process development strategy is well synchronized with internal production needs. The disadvantage, which sometimes may be considerable, is that group management, or top management in smaller organizations, will totally lack a good general overview of corporate innovation and that possible synergies between business or production units are lost. Well-designed corporate follow-up systems for innovation are one remedy for this disadvantage, and well selected and aggregable strategic corporate key ratios for innovation will diminish the risks of decentralized

R&D and can also be looked upon as an important part of a corporate "early warning system".

This final chapter of this book introduces the development of a corporate measuring system for innovation, of which process innovation is a substantial and important part in the process industries. In the spirit of the quotation from the above author, it is also hoped that the relevance of many of the previous parts of the book will be illuminated. The importance of establishing clear and well-functioning definitions and concepts for different kinds of innovation is highlighted as a way to clarify the charter for different parts of the R&D organization. Such well-defined concepts are not only essential to the development of good measurables, but also necessary to develop and implement the selected innovation strategies. Further on the importance of distinguishing between different degrees of "innovation newness", in the strategic choice of companies' project innovation portfolios, is also illuminated. Putting this information together on an aggregated corporate level will reinforce the strategic effectiveness and complement the more efficiency-oriented success factors. Such a "feed-forward" system of leading indicators will complement the feedback-oriented lagging performance indicators in the strategic learning process.

It is further advocated in this chapter that group management's follow-up of investments in fixed assets in the future ought to be supplemented by a similar and well-functioning system for follow-up of investments in intangible assets in general and innovation in particular.

16.1 Develop a measurement and follow-up system for innovation performance

The performance of individual production units, companies, divisions or business units in large corporations is usually assessed with the help of sophisticated financial information systems. R&D, R&D strategies and resource allocation to innovation are however normally left to the discretion of each profit centre to decide upon, and are therefore often not followed up at all at top management level. As stated before, there are several advantages to this order. These organizational units are closest to the external customers and have intimate knowledge of the needs of process innovation for technology and production. The drawback may on the other hand be that the necessary R&D, and especially long-term R&D, may be neglected in favour of cutting costs and maximizing short-term profits. If necessary resources are not properly allocated to R&D, such mismanagement does not unfortunately show up until after several years, when the damage is too late to repair. Referring again to the often long "time constant" for innovation, and for process innovation in the process industries in particular, the wrong distribution of allocated resources for innovation is equally harmful.

From the previous chapters it is evident that the question today is not so much whether it is possible to measure R&D, but how to design a proper corporate measurement and follow-up system for R&D. During this endeavour it will not hurt to bear the following simple questions in mind:

- What needs improving in innovation?
- What measuring instrument should we select?
- What units of measurement are appropriate?
- How often should we measure?
- Who is going to measure?

From a contingency perspective on corporate innovation activities (Woodward, 1965), well-functioning systems for innovation performance measurement must be adapted to each company's individual characteristics and the nature of its corporate innovation activities. It is recommended that each corporation should develop a customized system of corporate innovation strategy key ratios, success factors and performance measures for innovation, since there is no easily found off-the-shelf solution. The initiative to develop such a system is often of a "top-down" character, but the development of strategy, success factors and measurables must be carried out "bottom-up" within the

R&D organizations. In such a development, a stepwise work process assures organizational acceptance in a long-term improvement perspective. It is vitally important to measure relevant innovation outputs and success factors, but it is equally harmful to measure the wrong ones! Corporate success factors and performance measures must also in future be better presented and communicated to all corporate stakeholders, and the responsibility for such reporting rests solely on the R&D management.

The following text describes how to develop a customized system adapted to decentralized corporate innovation environments, a system which also may serve as an operational framework for all corporate levels.

In larger organizations, "pilot-testing" in a smaller organizational unit may be advantageous, but generally and if possible a simultaneous corporate roll-out is recommended, and a four-stage master plan for such an activity is given. Figure 13.1 shows the implementation of the system structured into four consecutive phases: prestudy, strategic intent, performance improvement and follow-up. These four main project phases recur in Figure 16.1, where the different parts of the individual phases are also shown.

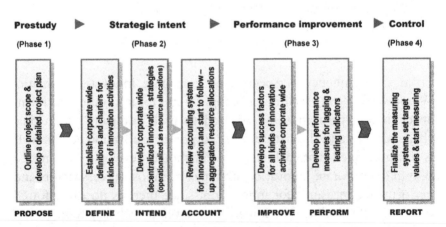

Figure 16.1   A four-stage master plan for system implementation. The four phases have been further split up into different parts, and the terminology is similar to the names given in the theoretical model presented in Chapter 13.

Prestudy (Phase 1)

Each company and corporation in the process industries operates in a unique industrial environment and market context. The prestudy therefore aims at identifying the specific corporate needs and demands on the development and implementation of such a system. The prestudy should be carried out under the supervision of one member of group management responsible for strategy or research and technical development or, in a smaller organization, the head of innovation. If somebody has been assigned corporate responsibility for management of innovation & technology as such, this person would be the ideal choice.

The first part of the prestudy focuses on fact-finding:

- Mapping corporate research and technical development functions, organizational affiliations and geographical locations.
- Collecting available information on how those activities are managed and controlled (informal and formal).
- Mapping and compiling the individual (different?) systems for resource allocation to innovation and the budgetary processes (if there is a yearly budgeting process).
- Compiling available data on the allocation and distribution of corporate resources for R&D.
- Present corporate state-of-the-art in the use of performance measuring of innovation.

The second part of the prestudy is pre-project desk work including:

- Putting together the above information (Report A).
- Preparing a draft project proposal for implementation of an innovation performance measurement system including objectives, delimitations and simplified time schedule, selecting a two-stage or simultaneous approach (Report B).
- Preparing an appendix that gives the necessary background information for innovation performance measurement and will serve forthcoming explanatory purposes (Report C).

- Presentation of the project proposal to group management to obtain clearance to take the proposal further and to develop a detailed project plan in collaboration with R&D management at different corporate levels.

The third part of the prestudy is to take this draft project proposal further to the various R&D organizations for discussions and further input and to secure support and collaboration for the proposed system. This can be looked upon as part of an improvement activity, since it will stimulate discussion and direct attention to this area of innovation management. The third step will include:

- Presentation and discussion of the project proposal in the various R&D organizations to secure support for the proposed system.
- Setting up a project organization and designing the detailed project plan for implementation.
- Making the final project official, communicating the information internally and planning the execution of the project in detail.

16.2 Transform the innovation strategy into key ratios (Phase 2) — you must walk the talk!

Establish (corporate-wide) definitions and charters for different kinds of innovation or innovation-related activities — DEFINE

A classification of innovation and innovation-related activities in the process industries appears in Chapter 3. The use of those definitions, which if necessary can be adapted and aligned with company traditions and specific needs, is now the backbone for the development of a sound corporate measurement and follow-up system for innovation.

Such a detailed definition of the innovation activities in the firm is something that must be developed as a first step on the road to system implementation. It is, however, generally regarded as a very rewarding exercise for the R&D organization, where many hitherto unasked questions about innovation charters and objectives are highlighted and answered. In the implementation it is sometimes important to stick to traditional corporate terminology so as not to disturb good familiar working practices.

Transform all the company's decentralized innovation strategies into aggregable defined key ratios for innovation — INTEND

Using the now clearly defined different kinds of innovation activities, key ratios for innovation will be developed, starting with the generally well accepted concept of innovation intensity.

R&D intensities — the overall key ratio

The innovation intensity is the starting point and the first kind of information to seek for (not forgetting the figure for absolute spending on innovation). The innovation intensity should reflect the overall strategic innovation intent in this business unit, subsidiary or production plant.

Distributed R&D intensities — key ratios for different kinds of innovation

Using the detailed differentiation of innovation activities presented in Chapter 5, the total annual expenditures for each area divided by the total spending on innovation are the distributed innovation intensities. For process innovation, for example, this is the percentage of total innovation. If some innovation-related activities are not included in the innovation budget, those figures could be presented as supplementary absolute figures.

Newness of innovation — key ratios for the newness of innovation

A further distribution and grouping of innovation activities is introduced for some areas of innovation. In Chapter 4, for example, the process innovation activity is further classified into several groups with the help of the process matrix, according to the newness of process development in two dimensions.

Review and start follow-up of allocation of innovation resources — ACCOUNT

After the above definitions and key ratios for innovation have been introduced, the figures are collected from the various R&D organizations. The implementation of the second phase of the system, strategic intent, is however not finished until the accounting system is also adapted to handle this information. This may include:

- Reviewing individual accounting systems and the corporate accounting systems for innovation and making them compatible with follow-up of the previously presented activities.
- Aggregating yearly expenditures from the lowest levels of R&D organizational structures to higher hierarchic levels (divisional and finally overall corporate level).
- Tracking results over time and making use of trend curves for the follow-up of yearly resource allocation to different areas of innovation.

It is generally considered by all parties a very interesting and rewarding process to apply these different key ratios and transform innovation strategies and budgets into comparable and aggregable figures for innovation. The aggregated figures at all levels, as well as at corporate level, will usually give food for thought. Do target values and budgets need a review?

16.3 Develop company-specific success factors and performance — measures (Phase 3)

Develop success factors for innovation — IMPROVE

Development of success factors is an exercise that must be carried out bottom-up in different R&D organizations, and the final compilation and selection of the important ones that will be used as corporate-wide improvement tools can be rather time-consuming. On the other hand, this must be considered as another part of the improvement process for innovation. These activities normally include:

- Generating potential success factors for different kinds of innovation — a truly bottom-up exercise.
- Compiling and refining the results.
- Internal benchmarking of the results.
- Selecting the final structure.
- Reviewing the results in all R&D organizations.
- Letting the individual organizations set their targets.
- Starting follow-up.
- Improving innovation performance and culture.

Develop performance measures — PERFORM

In the development and selection of key performance indicators it is important to involve and get as many research and development individuals as possible on board. The individual steps in this process are:

- Generating potential key performance indicators for different kinds of innovation — a truly bottom-up exercise.
- Compiling and refining the results.
- Internal benchmarking of the results.
- Selecting the final structure.
- Reviewing the results in the different organizations.
- Letting the individual organizations set their targets.
- Starting follow-up.
- Improving innovation performance and culture.

When a detailed proposal on both success factors and key performance indicators is worked out, it is generally recommended to get all personnel involved by rating and benchmarking the suggested success factors and key performance indicators in an internal survey.

16.4 Introduce a company follow-up system for innovation performance (Phase 4)

Follow-up — REPORT

The final phase of system implementation is related to how the information should be used and to which levels the information should be aggregated. This will include the following activities:

- Finalizing the measurement system.
- Designing the follow-up system.
- Compiling the target values of the R&D organizations.
- Starting measurements.
- Designing the reporting system — REPORT.

Some expected outcomes from the introduction of a follow-up system are listed below. The overall outcome can however simply be stated as an improvement of the company's management of innovation and technology.

- Establishing a common, well-defined language for innovation.
- A driver for development of improved decentralized innovation strategies.
- A driver for more efficient innovation and innovation work processes.
- An excellent tool for further internal and external benchmarking of innovation at different levels.
- Improving corporate management of innovation and technology in general and specifically the area of innovation performance measuring.

Companies that understand the importance of developing a measurement and follow-up system for innovation improvement that is founded on company-specific success factors and captures the firm's innovation strategy can look forward to improved future performance in innovation.

16.5 Discover and challenge the "hidden innovation strategy" in large decentralized environments

The conglomerate as a business model for multinational firms is out of fashion now, probably partly because such structures lack the ability to find business synergies and possibly also synergies in innovation. For process innovation in particular, there is a window of opportunity for larger firms to utilize individual process innovation results in a larger number of production plants. Many large corporations, however, are currently managed in a manner similar to that of the now abandoned pure conglomerates, and thus miss opportunities for innovation synergies and shared knowledge. Since there may be interesting opportunities for collaborative innovation activities in large corporations, one should thus not necessarily only get stuck on today's jargon "open innovation".

External collaboration under the banner of open innovation is naturally an important activity for all firms, but what about improved internal collaboration and communication in corporate innovation? The centre of large organizations, group management, may have different functions, ranging from an overall strong centralized planning role to mere financial control. The most common role today is however a more strategic control whose tasks include fostering innovation, defining standards and assessing and intervening to improve performance, as well as encouraging collaboration and co-ordination (Johnson and Scholes, 1999, p. 430). One parenting role for group management is thus to stimulate such internal collaborative activities. It is rather strange though, that in hard times when most corporate activities are measured and carefully evaluated in a short-term and long-term perspective, innovation often totally lacks a good follow-up system for strategic resource allocation and performance measurement at corporate level!

Is innovation, then, an exception to the old saying that what gets measured gets done? This is probably not the case, and since there seems, remarkably, to be an almost total consensus about the strong relationship between a company's future competitiveness and its innovation capability, the conclusion is that corporate innovation should not be left without a well-functioning follow-up system. The follow-up of innovation activities at corporate level is thus one tool that makes innovation activities more transparent in the corporation and could open up more efficient internal collaboration. Organizational forms that can support this mode of collaboration are not necessarily the old-established departmental structures, which instead often tend to create communication barriers, but more likely organizational solutions like the networks and virtual structures which have been discussed in Chapter 7.

It is not only important to develop good decentralized innovation strategies, which we could also look upon as innovation effectiveness (doing the right things). We must also consider innovation efficiency (doing things right). When combined imaginatively, these two areas make up good innovation productivity.

A shared follow-up system for aggregated strategic resource allocation for innovation at corporate level

Using overall innovation intensities and the further distributed innovation intensities compatible with the Strategic Research Agendas (SRAs) created at lower organizational levels, resources allocated to different areas of innovation can now also be further aggregated. From a plant level, the allocated resources can be further aggregated to subsidiary company level, and thence to division or business area, finally reaching overall corporate group level.

In a similar vein the newness of process innovation, for example, can be aggregated using the previously presented matrix for process innovation. This is illustrated in Figure 16.2; for each level, the distributed resources to different areas of the matrix are used to reach the final distributed newness at corporate level.

Since a company's progress is understood by looking back but has to be planned ahead, using the figures for the above indicators from the past, present and future can be an interesting point of departure for more factually based discussions of the articulated or "hidden" corporate R&D strategies. The trend curves over time for different organizational units' R&D intensities, together with the trend curves for distributed innovation intensities, will provide interesting background information for in-house discussions — even more so if the data are further supplemented with data from selected competitors.

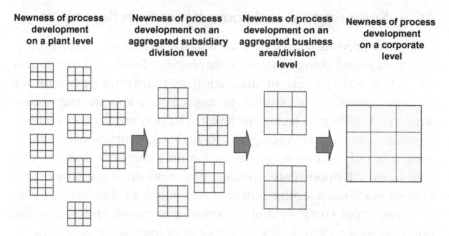

Figure 16.2 The aggregation of process innovation newness. The squares symbolize the "newness matrices for process innovation" shown in Chapter 4. Using the distribution of allocated innovation resources to the different areas of such matrices, the newness of process innovation, etc. can be aggregated to corporate level.

A common follow-up system for success factors and aggregated performance measures at corporate level

The advantage of overall and simultaneous development of a corporate measurement and follow-up system for success factors and performance measures is that they can be streamlined to suit the total corporate organization. The general development and agreement on individual success factors is then the first step on this road. Each organization can then, in an internal benchmarking exercise, further select those of highest importance to follow-up. The advantage of common output-oriented lagging measurables is evident, and development of such generally well accepted measures is not altogether too difficult to achieve.

Expected outcomes

- A tool to identify and share common corporate innovation best practices and core corporate values in innovation.
- A tool for improved cross-organizational dialogue and for discovering and utilizing corporate innovation synergies.
- Follow-up of corporate innovation strategies and an early warning system for corporate R&D.

16.6　Summing-up and some issues to reflect upon

Using the previously presented classification of innovation into different areas and using innovation intensities, distributed innovation intensities and newness of innovation as translating operators for strategic intent, it is possible to aggregate corporate innovation strategies to different hierarchic levels. The decentralized innovation strategies can thus be aggregated to higher organizational levels in order not only to follow up innovation on a corporate level, but to give an overall opportunity to streamline innovation initiatives. This kind of work has a strong potential to catalyse further iterative discussions across company and divisional boundaries and so has the potential to stimulate further discoveries of innovation synergies.

One objective of this book is to discuss innovation and innovation-related activities in order to create a stronger holistic corporate perspective on innovation. It is argued that innovation strategy-making is an iterative process and that the core of strategy-making should be close to external and internal customers for those related innovation activities. Nevertheless it is believed that corporate governance today should also address the question of innovation and innovation performance at corporate level. The message is not that strategy-making should be more top-down in the future, but that corporate group management also should follow up innovation activities at a group level. Such a follow up will also lead to an improved interaction, discussion and possible adjustments of such strategies and "hidden strategies". The outcome will be an improved and more explicit overall "group strategy" for innovation created in a bottom-up fashion.

Development of corporate generic success factors and performance indicators for innovation will not only provide an instrument for corporate follow up, but most importantly provide an excellent tool for further internal benchmarking and improvement of R&D performance. Assuming that setting standards for firm performance measurement is an important part of the group management's parenting role, it is recommended that setting standards for measuring innovation performance not should be neglected.

- *How often are innovation and innovation related topics discussed in board meetings?*
- *How often are innovation and innovation related topics discussed at the company or group level of your corporation?*
- *Has the performance of innovation ever been discussed during those meetings in general terms such as: "How good is our research, and how good is process innovation in particular?"*
- *Reviewing the expected outcomes on a group level in the last section, are they relevant to your corporation and if so, in what ranking order?*
- *Since investments in fixed assets usually are followed up at corporate level, what are the main reasons why intangible investments in innovation are not followed up in your organization at corporate level?*
- *Review the expected outcomes from the introduction of a follow-up system for innovation on a company level (Section 16.4), and sort them in ranking order for your firm!*
- *What is the performance improvement potential in your firm's innovation if innovation management tools that were more effective (e.g. better strategic choice of innovation areas and projects) and more efficient (e.g. better operational management tools and best practice) were implemented and better used? Use a scale between 0 percent and 100 percent.*
- *Review the suggested master plan for the introduction of a measurement and follow-up system for innovation performance. How well does such a plan suit your firm or how would you go ahead to introduce such a follow-up system in your firm?*
- *The proposed follow-up system includes the two parts strategic intent and performance improvements. Which one do you find of most interest to your firm?*

References

Adams, R, Bessant, J and Phelps, R (2006). Innovation management measurement: A review. *International Journal of Management Reviews*, 8, 21–47.

Balachandra, R and Friar, JH (1997). Factors for success in R&D projects and new product innovation: A conceptual framework. *IEEE Transaction on Engineering Management*, 44, 276–287.

Briner, W, Geddes, M and Hastings, C (1990). *Project Leadership*. Aldershot: Gower Publishing.

Brinkerhoff, RO and Dressler, DE (1990). *Productivity Measurement*. London: SAGE Publications.

Brown, MG and Svenson, RA (1988). Measuring R&D productivity. *Research Technology Management*, 31(4),11–15.

Chiesa, V, Frattini, F, Lazzarotti, V and Manzini, R (2009a). An exploratory study on R&D performance measurement practices: A survey of Italian R&D-intensive firms. *International Journal of Innovation Management*, 13, 65–104.

Chiesa, V, Frattini, F, Lazzarotti, V and Manzini, R (2009b). Performance measurement in R&D: exploring the interplay between measurement objectives, dimensions of performance and contextual factors. *R&D Management*, 39, 488–519.

Cooper, RG and Kleinschmidt, EJ (1995). Benchmarking the firm's critical success factors in new product development. *Journal of Product Innovation Management*, 12, 374–391.

De Saint-Exupéry, A (1984). *The Wisdom of the Sands*. Chicago, IL: University of Chicago Press.

Ellis, L (1997). *Evaluation of R&D Processes: Effectiveness Through Measurement*. Boston, MA: Artech House.

Eneroth, B (2005). *How Do You Measure Beautiful*. (in Swedish) Stockholm: NoK.

Foster, RN (1986). *Innovation — The Attacker's Advantage*. New York: Summit Books.

Foster, RN and Kaplan, S (2001). *Creative Destruction — From Built-to-Last to Built-to-Perform*. New York: Prentice Hall.

Galtung, J (1967). *Theory and Methods of Social Research*. Oslo: Universitetsförlaget.

Griffin, A and Page, AL (1991) PDMA Success measurement project: Recommended measures for product development success and failure. *Journal of Product Innovation Management*, 13, 478–496.

Hutcheson, P, Pearson, AW and Ball, DF (1996). Sources of technical innovation in the network of companies providing chemical process plant and equipment. *Research Policy*, 25, 25–41.

IRI (2006). *Measuring and Improving the Performance and Return on R&D*. Arlington, VA: Industrial Research Institute.

Johnson, G and Scholes, K (1999). *Exploring Corporate Strategy*. London: Prentice Hall Europe.

Kaplan, RS and Norton, DP (1992). The balanced scorecard: Measures that drive performance. *Harvard Business Review*, 70(1), 71–79.

Kaplan, RS and Norton, DP (1993). Putting the Balanced Scorecard to Work. *Harvard Business Review*, September–October.

Kaplan, RS and Norton, DP (1996a). *The Balanced Scorecard: Translating Strategy into Action*. Boston, MA: Harvard Business School Press.

Kaplan, RS and Norton, DP (1996b). Using the Balanced Scorecard as a Strategic Management System. *Harvard Business Review*, January–February.

Kerssens-van Drongelen, I (1999). *Systematic Design of R&D Performance Measurement Systems*. Twente: University of Twente.

Kerssens-van Drongelen, I, Nixon, B and Pearson, A (2000). Performance measurement in industrial R&D. *International Journal of Management Reviews*, 2, 111–145.

Knight, KE (1984). Technology transfer in the petroleum industry. *Journal of Technology Transfer*, 8, 27–34.

Lager, T (2001). Success factors and new conceptual models for the development of process technology in Process Industry. *Department of Business Administration and Social Science. Division of Industrial Organization*. Luleå, Luleå University of Technology.

Lager, T and Hörte, SÅ (2002). Success factors for improvement and innovation of process technology in process industry. *Integrated Manufacturing Systems*, 13, 158–164.

Lager, T and Hörte, SÅ (2005). Success factors for the development of process technology in process industry Part 1: a classification system for success factors and rating of success factors on a tactical level. *Int. J. Process Management and Benchmarking*, 1, 82–103.

Leonard-Barton, D (1992). The factory as a learning laboratory. *Sloan Management Review*, 34, 23–38.

Leonard-Barton, D (1995). *Wellsprings of Knowledge*. Boston, MA: Harvard Business School Press.

Leonard-Barton, D and Sinha, DK (1993). Developer-user interaction and user satisfaction in internal technology transfer. *Academy of Management Journal*, 36, 1125–1139.

Lindahl, O and Lindwall, L (1978). *Science and Best Practice.* (*In Swedish*), Malmö: Natur och Kultur.

Neelamkavil, F (1987). *Computer Simulation and Modelling.* Chichester: John Wiley & Sons.

Packer, MB (1983). Analyzing productivity in R&D organizations. *Research Technology Management.* Arlington, VA: Industrial Research Institute.

Robb, W (1991). How good is our research? *Research Technology Management,* 34(2), 16–21.

Roberts, EB (1995). Benchmarking the strategic management of technology — 2. *Research Technology Management,* 38, 44–56.

Samsonova, T, Buxmann, P and Gerteis, W (2009). Defining KPI sets for industrial research organizations — a performance measurement approach. *International Journal of Innovation Management,* 13, 157–176.

Schumann, PA, Ransley, DL and Prestwood, CL (1995). Measuring R&D Performance. *Research Technology Management,* 38, 45–54.

Tipping, JW (1993). Doing a lot more with a lot less. *Research Technology Management,* 36(5), 13–14.

Webster (1989). Webster's Encyclopedic Unabridged Dictionary of the English Language. New York: Portland House.

Werner, BM and Souder, WE (1997). Measuring R&D performance — State of the art. *Research Technology Management,* March–April, 34–42.

Virgil *The Aeneid.* London: Penguin Books.

Von Hippel, E (1988). *The Sources of Innovation.* New York: Oxford University Press.

Woodward, J (1965). *Industrial Organizations: Theory and Practice.* London: Oxford University Press.

Appendix A

Defining the Process Industries
by the NACE Classification System
(NACE, 2006)

B Mining and quarrying

05 Mining of coal
06 Extraction of crude petroleum and natural gas
07 Mining of metal ores
08 Other mining and quarrying

C Manufacturing

10 Manufacture of food products
11 Manufacture of beverages
17 Manufacture of paper and paper products
19 Manufacture of coke and refined petroleum products
20 Manufacture of chemicals and chemical products
21 Manufacture of basic pharmaceutical products and pharmaceutical preparations
22 Manufacture of rubber and plastics
23 Manufacture of other non-metallic mineral products
24 Manufacture of basic metals

D Electricity, gas, steam and air conditioning supply

35 Electricity, gas, steam and air conditioning supply

E Water supply, sewerage, waste management and remediation activities

Appendix B

The 2008 EU Industrial R&D Investment Scoreboard

Some definitions and methodology

The term "EU company" refers to a company whose ultimate parent has its registered office in a member state of the EU. The Scoreboard (Guevara *et al.*, 2008) uses the ICB (International Classification Benchmark) for the sectoral classification. The data for the Scoreboard are taken from companies' publicly available audited accounts. The Scoreboard refers to all R&D financed by a company from its own funds, regardless of where the R&D is performed (cash investment which is funded by the companies themselves). Definitions of R&D are according to the Frascati manual (OECD, 2002). Sales follow the usual accounting definition of sales, excluding sales taxes.

The sample

EU and non-EU groups include companies with different volumes of R&D investment. This year, the R&D investment threshold for the EU group is €4.27 million and for the non-EU group €24.21 million. If a comparison of EU and non-EU is sought on a similar basis, it is preferable to consider only EU companies with R&D above the non-EU threshold. This comprises a group of 402 EU companies, representing approximately 95 per cent of total R&D investment by the EU group. Using the non-EU threshold yields a sample of the world's top 1402 R&D investors that can be used for comparative purposes. Since the primary aim in this study is not to compare the EU with the rest of the world, all companies have been included to increase the sample size.

Some changes in the database for this book

The ICB database industry classification has been converted into the NACE classification system, using the definitions presented in the ICB database. Using the codes and the previously presented intensional definition of the process industries, the process industries have been divided into 11 "process industry sectors". The limited ICB statistical details prevent these sectors from being separated more usefully; for example the sector "chemicals" would have been better separated into "commodity" and "speciality" chemicals. In a similar vein the "food industry" includes farming and fishing because it is not possible to discriminate between them.

Appendix C

The European Research Project — Methodology and Sample

Unit of analysis, the population and the sample

The unit of analysis in this study is "company process development", and the sampling unit is the company. The level of analysis is sometimes the company or a higher sectorial level, but sometimes also the level of the whole process industry. The population is the entire group of objects about which information is sought. The unit (element) is an individual member of this population of which we take a sample using a selected sampling frame. Strictly speaking, the population for this study, as previously mentioned, is companies belonging to the category of process industry worldwide. For obvious reasons, the sample could not be a simple random sample (lacking any kind of sampling frame), so a number of industries were selected according to the following criteria:

Type of industry (old NACE codes)

Companies selected according to the previously presented definitions and the types of industries are also denominated using statistical codes for European industry (NACE, 1996). Industries from different sectors have been clustered together, and the number of companies in each sector is given after the NACE code. The sectors studied were:

- Mining and mineral (NACE codes 13, 14 and 26); 11 companies
- Food and beverage (NACE code 15); 17 companies

- Pulp and paper (NACE code 21); 9 companies
- Chemical — including petrochemical, plastic and rubber but not pharmaceutical (NACE codes 23, 24 and 25); 29 companies
- Basic metal (NACE code 27); 23 companies
- Other process industry (NACE codes 28, 37, 40, 41 and 24.4 plus some other industries connected to process industry); 23 companies

Geographical location

Most of the companies were selected from Sweden, and the total Swedish sample is 99 companies (not including 10 that declined), making it almost a census for Swedish process industry. A fairly large number of industries from other Nordic countries (Norway, Finland and Denmark) were also selected. The total sample of Nordic countries other than Sweden was 80 companies. A smaller sample of industries was selected from the rest of Europe (Germany, the United Kingdom, France, the Netherlands, Belgium, Italy, Austria and Switzerland), and the total sample of European countries other than Nordic countries was 148.

Size of industry and development intensity

The contacted companies had at least 200 employees and frequently many more than 500. Some parts of the industry sectors presented above were excluded from the sample because it was believed that the development intensity for this group was too low to be of interest; thus meat production and concrete production, for example, were excluded.

Summary

It is not feasible to sample process industry worldwide. The total sample for this study was 327 companies from European process industry focusing on Sweden. Although this is not a simple random sample, it is argued that it can be regarded as a fairly good and representative group of companies for process industry.

Research process and research methodology

The theoretical research process

Galtung (1967) discusses the role of time in the formation of hypotheses, and in an interesting discussion gives three alternative sequential models for the three following events:

- Formation of the hypothesis.
- Conditions in the world to which the data refer; the phenomena.
- Knowledge of the data by the investigator.

The traditional model of the research process is a process according to the order above, and the formation of the hypothesis is, in Galtung's words, "an educated guess without looking into the cards". Another alternative model starts with prevailing conditions in the world followed by data collection, and ends with the formulation of a hypothesis. One could describe this as looking into the cards before hypothesis formation, or that data are accounted for by the hypothesis. An argument given by Galtung for a hypothesis that is made to fit all data is "so much the better". The last model, starting with the researcher's phenomenological perception of the area to be researched (a sort of pre-understanding), followed by the collection of empirical data and finishing with conclusions and formation of a hypothesis, seems rather appropriate in an explorative study, and was consequently selected for this study.

Interviews with R&D managers

How do we tap into the complex dimension which we call "successful behaviour in process development"? Finding and selecting potential success factors for process development can be looked upon as sampling the universe of different independent variables that influence good process development performance, and as an analogy to sampling the population of process industries for this study. It was decided to interview a selected number of R&D managers from eight sectors of process industry to obtain a list of potential success factors.

The interviews were carried out in an interactive unstructured fashion and the respondents had previously been asked to select two finished process development projects, one "successful" and one "not so successful". The respondents were asked to present the two cases, and during the presentation to try to pinpoint factors of importance to the outcome of the project. This approach was considered to be more effective than simply asking the respondents to list a number of potential success factors or, as Yin (1994) puts it:

> "In some situations, you may even ask the respondent to propose his or her own insight into certain occurrences and may use such propositions as the bases for further inquiry. The more a respondent assists in this latter manner, the more the role may be considered one of an 'informant' rather than a respondent."

The author interacted during these presentations, asking questions and requesting the respondents to follow up or clarify certain points. The tape-recorded interviews normally lasted 2–4 hours and were often considered rewarding for both the interviewer and the respondents in gaining insight into process development. After the interviews had been completed, all potential success factors were written down on Post-it® Notes and were structured in clusters using an affinity technique. The clusters were then labelled and clustered together, creating a hierarchy of potential success factors in a bottom-up fashion. A three-level hierarchy of potential success factors was created as a platform for the development of the questionnaire part about success factors. To get a manageable number of success factors for the questionnaire, the secondary level with 25 potential success factors was selected for the respondents' rating, leaving the more explanatory bottom level for a supplementary ranking.

The measuring instrument — the survey

The Nordic group of companies received the questionnaire in Swedish and all others in English. The colour of the questionnaire was mauve, which did help the respondents to remember it when they were later reminded in the retrieval process. The questionnaire was pilot tested on three R&D managers from the previously mentioned interview group. All questionnaires were sent to a specific named person who had previously been identified as the right respondent in the organization. For practical reasons, the method of conducting the survey in Sweden differed from the one used in the rest of Europe; the main difference was that the Swedish respondents were contacted in advance by telephone before they received the questionnaire. The Swedish survey was carried out according to the following scheme:

- telephone contact with companies, checking the data and confirming participation;
- mailing the questionnaire;
- telefax reminders;
- telephone reminder;
- new telephone reminder;
- final telefax reminder.

The scheme for the Nordic and European survey was:

- mailing the questionnaire;
- new mailings after new information obtained (new contact persons or address);
- telefax reminder;
- final telefax reminder.

Appendix C — Table 1 shows the response rates from the survey. The response rate for Sweden is very good, considering that the total number of respondents, including those who declined to participate on the telephone before the mailing, is included in the sample.

For some sectors nearly all companies were included in the Swedish sample, making it close to a census. The response rate dropped dramatically for the rest of the sample because of the different design of the survey. The questionnaires were completed very well, with very few if any answers left blank.

Appendix C — Table 1   Response Rate for Different Geographical Areas in this Study

	Sweden	Other Nordic countries	Other European countries
Number contacted	109		
Number of mailings	99	80	148
Number of responses	78	18	16
Response rate	72%	23%	11%

A further discussion of methodological problems

Validity

As a general introduction to the concept of validity, let us consider the following formulation: "A variable is a valid measure of a property if it is relevant and appropriate as a representation of that property. Does the process measure what you want it to? To discuss the issue sensibly, we must ask validity for what purpose and validity for what population." (Moore, 1991). Validity is thus a matter of what the measuring instrument is really measuring. Usually one distinguishes between construct validity, internal validity and external validity.

Construct validity

Construct validity is used to describe how well the operational measure corresponds to the property one want to measure (the definition of that property). In this study great care was taken to define the concepts well. The overall important concepts "product and process development" and "success factors" were presented in great detail in the questionnaire. Understandability has been stressed and hopefully facilitated by a sometimes hands-on use of presented models. Success factors on a tactical level are fairly clear and well explained, while the success factors on an operational level are at times a little fuzzy. The R&D managers, as a group of respondents, must be considered as a very knowledgeable group of respondents for this type of questionnaire.

The way the questionnaires were completed gives no indication that the respondents had any difficulty in understanding the questions. The overall conclusion is that the construct validity of the study is high.

Internal validity

Internal validity is used in discussing causal relationships, which is the case in the area of rating and ranking "success factors", as previously discussed in the section on research variables. Considering the respondents' ability to correlate potential success factors to process development success, it is difficult to find more capable respondents for this task. The R&D Managers' ability to estimate the correlation between success factors for process development and process development performance must be considered to be high, but their ability to estimate the R&D function's share of a successful process development must be considered lower. Ideally a cross-functional panel of managers from different functions would have been a better alternative, but that was impossible to achieve in this survey. The internal validity is nevertheless considered to be high.

External validity — generalizations

External validity establishes the domain to which a study's findings can be generalized. There are two major types, statistical generalization and analytical or theoretical generalization.

Statistical generalization uses statistical means to make an inference from a sample to a larger population. A number of units are assumed to represent a larger population, so that data from a smaller number of units are assumed to represent the data that might have been collected from the entire population. This is normally used when the results are based on a survey using a design-based approach. The second system, analytical or theoretical generalization, was selected for use in this study. Here, research results are generalized to a model or a broader theory. This is the only type that is possible for case studies. It is also used in statistical inference and is then called a model-based approach, where the generalization is to a model or to a "superpopulation". The question of generalizing can also be related to the idea of finding the "representative case" (*le cas pure*) (Galtung, 1967).

This study is of an exploratory character even though a survey has been used here as the measuring instrument. Since a probability sample was not used (which in any case is very seldom possible in this type of study), statistical inference is not strictly applicable. Although no statistical generalizations are attempted in this study, it is convenient to discuss the companies selected in this study in statistical terms. Some formal statistical tests have been made, but it must be kept in mind that the statistical inference is not based on a firm probability sample to population ground. The results from rating of individual success factors show a remarkable similarity between different categories of process industry, which indicates that research results are usable for a larger group of industries, and are of a more generic quality.

There has been no intention to attempt a strict statistical generalisation of the findings to other process industries outside the group of industries participating in this survey, nor does this sample make any claim to be considered as a probability sample, so generalizations are from company process development to theory. The high response rate and share of Swedish process industries (the participation from some sectors is close to a census) makes the validity high for Swedish process industry but rather poor for European process industry. It could however be argued that geographical location ought to have less influence on success factors for process development than the nature of the development work. How true are the results? The theoretical model of the research process presented earlier shows that the results have sometimes given support to the formulation of a number of unverified hypotheses.

Reliability

Reliability is about repeatability and lack of measurement errors in the measuring instrument and measurement process. A result cannot be valid if it is not reliable. In this study the R&D managers were selected as the respondents and the representatives of company R&D, but it could be argued that other people, in or outside the R&D organization, should have been selected instead. It would probably have been better if the questionnaire had been completed by a group of company representatives since some questions call for reflection and discussion.

In that case there could then have been several alternatives. Options could have been to select representatives from top management, or from the R&D management; and some questions could have been best answered by the staff of the R&D organization. None of these solutions was deemed feasible for this study because they would have made excessive demands on company resources. Nevertheless, it must be kept in mind that the results from the survey are the voice of the company R&D managers alone.

The reliability of the research results from this study may vary; some results must be considered very reliable and others not reliable. For example, the figures given in the "process matrix" for how the process development resources distribute to different areas of process development must be used with care, as it can be assumed that some respondents did not have the exact figures available and that the matrix was a new one. The rating and ranking figures are of a kind that could possibly differ if the same questions had been asked again after half a year (the stability aspect of reliability). There could also be a risk that the respondents, after working through the questionnaire for some time, could have been slightly biased by only getting questions about process development, and the importance of process development compared to other R&D activities might therefore have been overemphasized. The overall impression from the results is, however, that the reliability is acceptable for the purpose of this exploratory study.

References

Galtung, J (1967). *Theory and methods of social research*. Oslo: Universitetsförlaget.

Guevara, HH, Tübke, A and Brandsma, A (2008). The 2008 EU Industrial R&D Investment Scoreboard. Luxembourg: Office for Official Publications of the European Communities. http://iri.jrc.ec.europa.eu.

Moore, DS (1991). *Statistics — Concepts and Controversies.* New York: WH Freeman and Co.

NACE (1996). Statistical classification of economic activities in the European Community Rev 1. Luxembourg: Office for Official Publications of the European Communities.

NACE (2006). Statistical classification of economic activities in the European Community Rev 2. Luxembourg: Office for Official Publications of the European Communities.

OECD (2002). *Frascati Manual: Proposed Standard Practice for Surveys on Research and Experimental Development.* OECD Publishing.

Yin, RK (1994). *Case Study Research; Design and Methods,* Thousand Oaks, CA: Sage Publications.

Index